BEYOND MECHANISM

The Universe in Recent Physics and Catholic Thought

Edited by
David L. Schindler

Contributors:

David Bohm
John B. Cobb, Jr.
Frederick J. Crosson
William J. Hill
David L. Schindler
Kenneth L. Schmitz
John H. Wright, S.J.

UNIVERSITY
PRESS OF
AMERICA

LANHAM • NEW YORK • LONDON

Copyright © 1986 by

University Press of America,® Inc.

4720 Boston Way
Lanham, MD 20706

3 Henrietta Street
London WC2E 8LU England

All rights reserved

Printed in the United States of America

Co-published by arrangement with
Communio International Catholic Review
Library of Congress Cataloging in Publication Data

Beyond mechanism.

The present collection has their origin in a conference sponsored by the North American edition of Communio International Catholic review, held at the University of Notre Dame, 1984.
  Contents: Introduction : the problem of mechanism / David L. Schindler — The implicate order : a new approach to the nature of reality / David Bohm — Bohm's challenge to faith in our time / John B. Cobb, Jr. — [etc.]
  1. Cosmology—Congresses. 2. Physics—Congresses. 3. Mechanism (Philosophy)—Congresses. 4. Religion and science—1946-   —Congresses. 5. Bohm, David—Congresses. I. Schindler, David L., 1943-   . II. Bohm, David. III. Communio (Spokane, Wash.)
BD511.B47 1986   113   86-7787
ISBN 0-8191-5357-5 (alk. paper)
ISBN 0-8191-5358-3 (pbk. : alk. paper)

All University Press of America books are produced on acid-free paper which exceeds the minimum standards set by the National Historical Publications and Records Commission.

ACKNOWLEDGEMENTS

This book is published under the auspices of the North American edition of *Communio International Catholic Review*. The conference was sponsored by *Communio*. in association with the Center for Continuing Education of the University of Notre Dame, and with support from the Center for Process Studies, Claremont, California, and the Program of Liberal Studies, the College of Arts and Letters, and the Office of the Provost of the University of Notre Dame. Public thanks are herewith given for this support.

CONTENTS

List of Contributors vii

Preface ix

David L. Schindler: Introduction:
    The Problem of Mechanism 1

David Bohm: The Implicate Order: A New
    Approach to the Nature of Reality 13

John B. Cobb, Jr.: Bohm's Challenge to
    Faith in Our Time 38

Frederick J. Crosson: Man and the Meaning
    of the Whole 51

John H. Wright: Cosmic and Human Evolution
    in Theological Perspective 65

William J. Hill: The Implicate World: God's Oneness
    with Mankind as a Mediated Immediacy 78

Kenneth L. Schmitz: Temporal Integrity, Eternity,
    and the Implicate Order 99

David Bohm: Comments on the Papers 128

David Bohm: Hidden Variables and the
    Implicate Order 144

LIST OF CONTRIBUTORS

David Bohm, emeritus professor of theoretical physics, Birkbeck College, University of London. Books include *Quantum Theory, Causality and Chance in Modern Physics, Quanta and Reality*, and *Wholeness and the Implicate Order*.

John B. Cobb, Jr., Avery Professor of Religion, Claremont Graduate School, Ingraham Professor of Theology, School of Theology at Claremont, and Director of the Center for Process Studies, Claremont. Author of twelve books, including *A Christian Natural Theology, The Christian Structure of Existence*, and *The Liberation of Life: From the Cell to the Community* (with Charles Birch), and co-editor, *Mind in Nature*.

Frederick J. Crosson, Cavanaugh Professor in the Humanities, Program of Liberal Studies, University of Notre Dame. Former editor of the *Review of Politics*, editor of *Human and Artificial Intelligence* and *Contemporary Society*, and author of numerous articles in philosophy.

William J. Hill, O.P., professor of systematic theology, The Catholic University of America. Past president of the Catholic Theological Society of America, editor-in-chief of *The Thomist* (1975-83), and on the editorial board of *New Catholic Encyclopedia, Listening*, and the North American edition of *Communio* (until 1983). Publications include *Knowing the Unknown God* and *The Three-Personed God*.

David L. Schindler, associate professor, Program of Liberal Studies, University of Notre Dame, and editor-in-chief of the North American edition of *Communio*. Co-editor of *Act and Agent The Philosophical Foundations of Moral Education*, and the author of numerous articles in philosophy and theology.

Kenneth L. Schmitz, professor of philosophy, Trinity College, University of Toronto. Member of the Board of Editors of the North American edition of *Communio* and past president of the American Catholic Philosophical Association, the Metaphysical Society of America, and the Hegel Society of America. Publications include *The Gift: Creation*.

John H. Wright, S.J., professor of systematic theology, Jesuit School of Theology at Berkeley and the Graduate Theological Union. Member of the Board of Editors of the North American edition of *Communio*, and past president of the Catholic Theological Society of America. Author of *The Order of the Universe in the Theology of St. Thomas Aquinas*.

PREFACE

The papers which make up the present collection have their origin in a conference entitled, "Beyond Mechanism: The Universe in Recent Physics and Catholic Thought." This conference was sponsored by the North American edition of *Communio· International Catholic Review*, and was held at the University of Notre Dame in the spring of 1984. The theme which occasioned the conference and served as its general backdrop is developed in the introduction. Simply stated, the purpose of the conference was to reflect on the meaning of nature *(physics)*, in light of some of the central concerns of Catholic theology and philosophy. The immediate context for this reflection was set by the developments in twentieth century physics, particularly as interpreted in the work of theoretical physicist David Bohm. The papers give expression to the dual focus of the conference: some develop the conceptual issues more directly on their own terms (Wright, Schmitz, Hill); some develop the issues more directly in terms of the work of Bohm (Cobb, Crosson). In addition to an opening paper in which he sets out the mainlines of his interpretation of contemporary physics, Professor Bohm provides comments on each of the other papers, as well as an account of how he arrived at his own views.

Communio is grateful to the following persons who participated and greatly enriched the discussions of these papers at the conference: James Cushing, Thomas Langan, Ernan McMullin, Val J. Peter, John R. Sheets, Mary Katherine Tillman, Philip Sloan, Walter Nicgorski, W. Norris Clarke, and William Wallace.

I wish to express my gratitude to Sonja K. Jordan, Executive Editor of *Communio* for assisting in the final preparation of these papers.

David L. Schindler
University of Notre Dame

INTRODUCTION: THE PROBLEM OF MECHANISM

David L. Schindler

The aim of the conference from which the papers in this collection are drawn was to inquire into and elucidate the fundamental concepts in terms of which the universe is understood. Such a statement of the aim of the conference carries a range of meanings, especially in view of the title of the conference, "Beyond Mechanism: The Universe in Recent Physics and Catholic Thought." This title indicates that the inquiry to be engaged was both focussed in terms of mechanism seen as a problem and carried on jointly from within the perspectives of physics, philosophy, and theology. But engaging an inquiry of this sort carries with it a number of presuppositions, and it is therefore my purpose in this introduction to try to bring some of those presuppositions into relief.

I

(1) The central meaning of mechanism on the present context I think can best be reached by looking briefly at two fundamentally different ways in which the terms *physis* (nature) and physics have been understood in the history of Western thought. The first of these is represented in Aristotle, in what may be called the classical Greek view. The second is represented in Descartes, in what may be called the classical modern view. I will consider each of these views in turn.

The term physics, comes from the Greek *physis* and thus in turn from the verb *phyo,* which means "to bring forth, produce, put forth; to beget, engender, generate" and so on (Lidell and Scott *Lexicon).* And *physis* is rendered into English as "nature," which comes from the Latin *"natura,"* and thus in turn from the verb *"nascor"* which, like *"phyo,"* is also associated with giving birth. We can see developed in Aristotle the full meaning of these initial etymological considerations. I will focus that development in terms of a concept which figures centrally in any understanding of physics, namely, the concept of matter. How does Aristotle conceive nature *(physis)* relative to matter *(hule)?*

The answer I believe is essentially this: in one sense nature is matter: "Nature means the primary

material--*ex ou protou*"--of which any natural object consists or out of which it is made...."
(*Metaph*ysics 5.4. 1014^b 27-28; cf. also *Physics*, 2.). Nonetheless, in a second and indeed more proper sense, nature for Aristotle is *not* matter: for nature in the full and proper sense for Aristotle is something actual and matter in its basic meaning is not actual--it is what has the capacity for becoming actual. It follows that nature in its full and proper sense must be--not matter--but that--the act or activity--in virtue of which nature, and hence matter, are said to *be actual*. What I wish to suggest here, then, is this: that the meaning of *physis* (nature), is for Aristotle disclosed in the first instance by the meaning of act (that in virtue of which something is said to be actual). I thereby wish to suggest that the meaning of matter, in the sense of matter *as it is a part of nature in its proper--actual--sense,* is likewise disclosed in the first instance by the meaning of act. Thus it becomes crucial to get clear about what Aristotle means by act, and what he considers the main features of actuality.

There can be no question here of following Aristotle in all the nuances of his understanding as it bears on this issue. It will suffice to call attention to several of the terms which Aristotle employs as he unfolds the meaning of *physis* in his *Physics* and *Metaphysics*. The terms are these: *ousia* (being; beingness; what most fundamentally is: Ross translates this in the most relevant passages under concern here as essence. In any case, however, one translates the word, it seems to be crucial always to keep in mind that the words is a derivative form of the verb "to be" (*<eimi>*)· *arxe, enuparxo* (source; immanent source--Ross's translation); *sumphyo* as distinct from *aphes* (that is, growing together--what Ross translates as organic unity, --as distinct from contact or merely touching); *morphe* and *eidos* (shape and form); *telos* (end or finality--from *finis,* the Latin translation of *telos); entelexeia* (actuality, that is, that which, in constrast to *dynamis,* has completeness and perfection (*per-facio:* done all the way through, hence whole); *energeia* (actuality or activity).

A vast array of issues is of course introduced here by these terms and the translations I have offered. My purpose in introducing them is simply to suggest the range of what is involved for Aristotle in understanding *physis* (nature) in its full and primary sense, which is to say, in its actuality. It is thereby to suggest what is involved in understanding that from or in relation to which matter takes its meaning--its meaning as actual. Thus my suggestion is that for Aristotle, proper understanding of nature, of matter as an actual part of nature, requires an understanding of what it means to be in the most fundamental sense *(ousia)*, and this in turn is seen to require understanding of such features as immanence (immanent source of the activity in terms of which something is said to be actual, to grow, to change, to move), form (act of forming), end (act of finalizing), actuality or activity, and completeness or wholeness. Further, proper understanding of nature requires an understanding of the distinction between growing together as a unit and relating merely by contact.

Again, the complexity of the issues introduced here is enormous. Nonetheless, these brief remarks will suffice to indicate, relative to the theme of the conference, what I wish to propose is central to Aristotle's understanding of *physis:* namely, (a) that matter is a *relative* concept; it is something which can properly be understood in its actuality, only and always *in relation*--to nature in a fuller, proper sense; secondly, (b) nature, that *in relation to which* matter takes its full meaning, is characterized by act or activity which is immanent, formal, final, unifying, and complete or whole. This understanding of *physis* has often been called organic or organismic. These terms seem to me apt, since they are commonly taken to be characteristic of organisms i.e., the immanent activity of form and finality, internality of relation among the distinct "parts" of the organism, and consequently a wholeness of the organism which is distinct from the sum of its "parts. I thereby offer them as a summary of the features required, on an Aristotelian reading, for understanding matter properly, in any of its actual instances.

(2) When we turn to consider Descartes, the essential difference of hsi understanding of *physis* from that of Aristotle seems to me readily apparent. To put that difference simply, the matter which for Aristotle is a relative concept becomes for Descartes an absolute concept. In other words, the matter which (in any of its actual instances) is understood by Aristotle only (always and already) in relation to a nature whose primary act is formal and final becomes in Descartes precisely identical to a nature from which mind (anything like forming and finalizing activity) has--always and already--been removed.

The heart of Descartes's understanding of matter, then, unfolds like this: matter is something which can be clearly distinguished from mind. But what can be clearly distinguished for Descartes is what can be clearly set off from everything else (cf. *Principles of Philosophy*, Part I,; 45), and the best way to accomplish such a setting off of one thing from another is to "picture them in the form of lines" *(Discourse, Part 2)*. Thus, matter and mind (that is, anything like the immanent activity of forming and finalizing) are understood by Descartes as distinct in the way in which two things laying on opposite sides of the line from each other are distinct: that is, as both separate and simply different from each other, and thus as completely apart from each other. The content of the difference between matter and mind for Descartes consists in the fact that matter is what is extended.

The upshot of this reasoning by Descartes, then, is this: that the meaning of *physis* (nature) is now absorbed into a matter understood as that from which everything but extension has already and in principle been removed. There are at least two crucial transformations in the understanding of *physis* which occur here. First of all, nature (matter in its actual instances) seen by Aristotle to be internally active (e.g., formal and final) now becomes a nature (matter) without such internal activity, hence a nature (matter) which is essentially inert (or as Descartes says, in repose). To put another way, any activity to be ascribed to nature (matter) must now be of an external sort.[1]

Secondly, Descartes's understanding of *physis* (nature) entails a profound transformation in terms of how one might now characterize natures in any of its actual instances as a whole. For nature, now become a matter from which anything "more" like mind--forming and finalizing activity--has already been excluded, and thereby, can be a whole only after the manner in which what is exhaustively extension is whole, that is, precisely as quantity--a quantified bit or the sum of quantified bits. In other words, the wholeness of nature (in any of its actual instances) which for Aristotle is characterized as an internally active unity (and hence as a wholeness which is always, in principle, more than the sum of its "parts") becomes in Descartes a wholeness best characterized as an externally interactive collectivity (and hence as a wholeness which is always, in principle, exactly the sum of its parts).

4

To summarize, then: before we called Aristotle's understanding of *physic* organic or organismic, and suggested that the heart of that understanding consisted in a twofold claim: (a) that matter is a relative concept, something whose meaning (in any of its actual instances) is disclosed only in relation to nature in its fuller, more proper sense; and (b) that nature, that in relation to which matter takes its meaning (as actual) is (in each of its instances) whole (a unity which is more than the sum of its "parts") and is internally active. I now propose that we call mechanical or mechanistic Descartes's understanding of *physis*. I suggest that the heart of that understanding consists in a twofold claim: (a) that matter is an absolute concept, something apart from, not-relative to, anything "more" like internal--formal and final--activity; (b) that nature, now absorbed into matter in this way, is whole (in any of its instances) only in the sense of being a collection which is exactly the sum of externally interactive parts.[2]

Now it is important to note that the positions of Aristotle and Descartes as briefly described here are offered only as examples--albeit classical ones--of what it means to understand *physis* respectively as organic or mechanical. There are, of course, countless variations of each of these understandings. Thus, for example, with respect to the organic view of *physis*, one might argue, in the context of the Christian tradition, that Aquinas's understanding of act in its primary sense as *esse*, and indeed in its ultimacy as the *Esse* he calls God, entails some corresponding revision of the Aristotelian understanding of actuality--and hence some corresponding revision in the meaning of the whole range of terms which Aristotle involves in his understanding of nature: *ousia; arxe; sumphyo; eidos; telos; entelexeia; energeia,* and so on. And a revision in the meaning of these terms, and thus in the terms relative to which matter takes its meaning (in any of its actual instances), would seem further thereby to entail just so far a revision in what one ultimately means by matter.

But the point I should like to make here apropos of the organic understanding of *physis* which I have illustrated with reference to Aristotle is simply this: that it is the concept of matter as relative to a distinct (but internal) act (activity) for its meaning (as actual) which establishes one's understanding of *physis* as organic. Thus it follows that what some would argue is a distinct understanding of act in

Aquinas (from that of Aristotle) leaves intact the view of nature which would thereby qualify Aquinas's view *as organic*, all the while that it changes (proportionately to the difference in the understanding of act) the fuller meaning of that nature as organically conceived.

On the other hand, then, one might call attention to the numerous variations in what we have exemplified in Descartes as a mechanical or mechanistic understanding of *physis*. First of all, there has of course been an enormous refinement in the mechanistic understanding of nature (matter) as extension or quantity: Descartes's matter as inert substance undergoing change of place has become an enormous variety of micro-substances, as it were (elementary particles), undergoing change of place. More importantly, whereas Descartes's mechanism made the explicitly metaphysical claim that material entitites *really were* (in themselves) simply mechanical in their activity, a more methodological mechanism would abstract from the question about what material entities really are (in themselves), and restrict itself rather to treating those entities *as if* they were mechanical in their activity--treating them, that is, just so far as they manifest in mechanical ways. Finally, then, this more methodological mechanism is now of a statistical sort: it treats material entities in so far as they manifest in terms of mechanical order--particles, aggregates of particles--and this is seen to be an occurrence of statistical frequency.

In short, what I have described in relation to Descartes as the mechanistic understanding of *physis* can be and indeed has been conceived in ways quite different from that of Descartes. Nonetheless, the point I would want to make here is that that understanding remains mechanistic in the crucial respect just so far as it continues to treat matter only to the extent that matter manifests as discrete quantity (even if infinitesimally small and accessible as such only in terms of statistical frenquency). One's understanding of physics remains mechanistic just so far as one thus continues to treat matter as absolute, in the precise sense of "something" which is taken to be properly understood *in abstraction from* (non-relatively to) anything "more" such as internal activity (e.g., form and finality).

II

Now one of the most obvious things to be said about the mechanistic understanding of *physis* as I have described it is that it has been enormously successful. One of the hallmarks of modern Western culture is surely the remarkable achievements in medicine and technology which have their historical roots in the mechanistic understanding of physics dating from the seventeenth century. Nonetheless, the title of the conference suggests, in the face of these achievements, that mechanism is a problem, and that there is need to get beyond it. What does this mean?

The sense in which mechanism is understood to be a problem, and accordingly the sense in which there is seen to be a need to get beyond mechanism, is suggested by noting what seems to me to be a second obvious fact about the mechanistic understanding of physics: namely, that it has been indissolubly linked with what one might call the fragmentation in modern Western patterns of thought and life. Use of the term fragmentation of course carries a presuuposition. For the root meaning of the word *(frangere,* to break) denotes what is broken. And what is broken is incomplete: it is something which is detached in some inappropriate way. Or, to put it in other terms, a fragment might be called a part which is detached from the whole when and in ways that it ought not to be. A fragment is a part which is treated as separate from <exclusive of relation to> the whole when its very nature as part is properly understood rather to be inclusive of relation to the whole. To get at the senses in which what I have called the fragmentation of modern Western culture was a concern of the conference, I turn now to consider how the mechanistic understanding of *physis* has served to shape, not only--of course--the content and method of physics itself, but also the content and method of philosophy and theology. I will focus was what seems to me to be three important problem areas: the nature of immateriality (non-material reality; spirit), the nature of value, and the nature of metaphysics and theology as matters of knowledge.

But first a general comment. The general comment consists in calling attention at the outset to the difference in the contexts in which Aristotle and Descartes work out what I have offered as classical formulations of the organic and the mechanical understandings of *physis* (nature). Descartes's understanding of nature is developed by him in what is, at least as a matter of explicit concern, an anthropological context. In other words, the meaning of nature (matter) is reached by him in the first instance in the context of (albeit by contrast with) what he takes to be specifically human activity, namely thinking. For Aristotle on the other hand, nature can be properly understood only by considering the meaning of activity in the first instance, not in term of specifically *human* being, but rather, more basically, in terms of *being (ousia)*. The proper context for understanding nature thus for Aristotle is not in the first instance anthropological (as for Descartes), but metaphysical and indeed theological (the nature of the activity of being qua ultimate).

This is of course a profoundly difficult issue, and reasons can be offered for Descartes's shifting (by way of restriction) of context. My purpose here is simply to record that difference in context and to underscore its significance for the subsequent modern understanding and discussion of the problems of nature/ matter: quite simply, in an Aristotelian context one would consider the problem of nature in terms of the meaning of matter relative to activity--act, actuality--in its most fundamental sense *(ousia)*; in a post-Cartesian context one tends on the contrary to consider the problem of nature--in the first instance--in terms of the meaning of matter relative to activity in its peculiarly human sense. I believe the significance of this shift in context is felt in each of the three problem areas I now wish to focus.

(1) First of all, then, the mechanistic view of *physis* has affected our understanding and discussion of anything--any agent or agency--we might take to be immaterial (non-material or indeed spiritual): form, finality, mind, God. The heart of the difference in the understanding of immateriality carried in mechanism I suggest lies in the loss of immanence as the central feature of immateriality, a loss which can be seen in at least three ways. The fundamental sense of the loss of immanence is indicated simply in recalling the way Descartes first reaches his understanding of mind (and consequently immaterial agency): that is, by his distinguishing mind after the manner of what can be set off from something else after the fashion of what is on the opposing side of the line from that something. What this means is that mind (immaterial agency) is just so far conceived in the first instance in the spatial terms of what is *outside* or *external*. The

point is crucial: for Descartes's understanding of *physis* thus does not only *result* in the elimination of immanence as a feature of immaterial agency; it *is* the elimination of immanence.

Of course it is possible still to conceive the mind, as, Descartes in fact does, in some sense inside matter. But note how--and this is my second point relative to the loss of immanence--the meaning of "inside" (inner, internal) now gets transformed. The mind immaterial agency is (can be) inside matter (the body) only after the manner of what is disjoined or separate (somewhat after the manner in which we might picture something like a gremlin lurking at the center of a machine: not *in*-- immanent within--the machine; but rather remaining external to the machine albeit now from somewhere imagined to be its center). It is not hard to see how, subsequently, the activity of the mind (and indeed that of any agent taken to be immaterial: e.g., God) comes to be understood as something private: that is, as something hidden by rather than disclosed *in* the matter (bodies) taken to be public. And further it is not difficult to see how the activity of the mind (and indeed that of any agent taken to be immaterial: e.g., God) subsequently comes to be understood as little more than an arbitrary intrusion on matter (bodies).

Thirdly, and in connection with these two statements, it follows that, in so far as one does continue to affirm anything like a distinct activity on the part of any immaterial agent (mind, God) on nature (matter), that activity can just so far be conceived only in terms of what comes simply from outside: hence in terms of the sort of activity from without which we commonly call forceful.[3]

In these three senses, then, I suggest that the mechanistic understanding of *physis* profoundly changes our understanding of immateriality. It seems to me unnecessary to rehearse all the ways in which such an understanding of immateriality (of anything taken to be immaterial--or spiritual) has in recent centuries in the West been operative in discussions regarding the meaning of mind (mind acting relative to body) and of God (God acting relative to the world).

(2) I turn next to a consideration of how the mechanistic understanding of *physis* has affected our understanding of the nature of value--and hence of moral, esthetic, and religious values. I begin by

assuming that the nature of value is indissolubly linked with something like the immanent activity we have called form and finality. Taking this to be the case, then, it follows that the mechanistic view which consists in removing such activity from nature, and in so far as it does, just so far entails removing value from nature. Values are thus just so far seen to be no longer rooted in nature, hence just so far to be non-natural. In such a context, it is inevitable--and Western history bears this out--that values come to be seen--that is, because non-natural--just so far as arbitrary. Further, because they are no longer disclosed *in* nature and hence are not accessible in the public way that nature (matter) might be said to be accessible, it is likewise inevitable that values come to be seen as essentially private matters. Once again I do not think it is necessary to rehearse all the profound ways in which this separation of nature and value (commonly called a dualism of fact and value) has figured prominently in modern Western patterns of culture.

(3) Finally, let us note briefly how the mechanistic understanding of *physis* (nature) has in recent centuries affected the meaning of meaphysics and theology as matters of knowledge. The point can be made simply. The mechanistic understanding of nature, which entails a disjunction between nature (matter) and what is not nature, entails in turn a disjunction between the sort of knowledge proper to the study of nature (e.g., physics) and the sort of knowledge proper to the study of what it would take to be non-nature (e.g., what one might call metaphysics and theology). But once one conceives these studies (and their methods) in this disjointed fashion, two important consequences are likely to follow. On the one hand, metaphysics and theology come to be viewed as knowledge of a simply derivative sort (simply derived from physics), that is, in so far they continue to be seen as concerning knowledge of nature at all. On the other hand, in so far as metaphysics and theology are not seen any longer to concern knowledge of nature, they come to be viewed as not properly matters of knowledge at all (i.e., because not knowledge in its natural form). Again, I do not think I need rehearse here all the significant ways in which modern patterns of thought have reflected this view of metaphysics and theology.[4]

I focus all these problem areas, not of course with the intention of suggesting that the issues raised therein are susceptible of any easy resolution. There are countless variations and more subtle forms of the positions which I have only sketched, and there are reasons which might be marshalled in defense of each of these positions. My purpose in outlining the positions is only to suggest important ways in which it seems to me that mechanism has manifested itself as a problem. I have offered three such ways, bearing respectively on the meaning of immateriality (hence spirit), value, and metaphysics and theology. My intention has been to show, by means of these examples, not only how the mechanistic understanding of nature transforms the meaning of *physis* --and hence bears on the work of physicists; but also to show how, in so doing, that mechanistic understanding simultaneously and profoundly affects a whole range of other concerns: one's conception of mind, of God of moral and esthetic and religious values, of metaphysics and theology. My purpose has thus been to show that the mechanistic understanding of nature is also a matter of profound relevance to the work of philosophers and theologians.[5]

INTRODUCTION NOTES

¹This activity is of course often called effective or efficient. I would only note that the elimination of internality--form and finality--entails a profound change from the Aristotelian understanding of efficient cause or activity: the latter is now reduced to only one of its possible meanings as found in Aristotle. And again, movement for Descartes becomes local movement or the displacement of bodies: "the transference of one part of matter or one body from the vicinity of those bodies that are in immediate contact with it, and which we regard as in repose, into the vicinity of others." It is extremely interesting and important in this context to note the difference in meaning between Aristotle's *kinesis*, usually translated as change or movement, and what Descartes means here by movement.

²Note the aptness here of the term mechanical/mechanistic. The term comes from the Greek verb *mexanaomai*, which means to contrive or make a device of some sort. The key then is the notion of something which comes together arbitrarily in the sense that its "parts" are already complete apart from such coming together--hence the relation between them in the device/contrivance can only be external.

³Cf. my earlier comment on p. 7 above regarding the meaning of efficient or effective activity in its differences as found in Aristotle and Descartes.

⁴It is interesting of course in this context to note how physics--in its mechanistic understanding--came to be called science (that is, *scientia*, knowledge): physics and its method--of observation/quanitification--become the only way to acquire positive knowledge in the proper sense, all other endeavors--such as philosophy and theology--becoming at best second order/derivative enterprises.

⁵For further exploration of this theme, cf. my "Beyond Mechanism: Physics and Catholic Theology," *Communio: International Catholic Review*, Vol. XI, No. 2 (Summer, 1984) pp. 186-192.

THE IMPLICATE ORDER: A NEW APPROACH TO THE NATURE OF REALITY

David Bohm

1. *Two World Views, Organicism and Mechanism*

Throughout history, there has been a succession of world views, i.e., general notions of cosmic order and of the nature of reality as a whole. Each of these views has expressed the essential spirit of its time, and each of them in turn has had profound effects on the individual and on society as a whole, not only physically, but also psychologically and ethically. These effects were multiple in nature, but among them one of the most significant is that of notions of universal order.

I shall begin by giving two examples of such world views that are of key importance for this discussion. The first of these is the ancient Greek notion of the Earth at the center of the universe, with seven concentric spheres in the Heavens, in an order of increasing perfection of their natures. Together with the Earth, they comprised a totality that was considered to be an integral organism, whose activities were regarded as meaningful. As suggested especially by Aristotle, each part then had its proper place in this organism, and its activity was seen as an effort to move toward this proper place, and to carry out its appropriate function. Man was thought to be of central importance in the whole system, and this implied that his proper behavior was to be regarded as correspondingly necessary for the over-all harmony of the universe.

In contrast, in the modern view, the Earth is a mere grain of dust in an immense universe of material bodies (stars, galaxies, etc.). These, in turn, are ultimately constituted of atoms, molecules and structures built out of them, as if they were parts of a universal machine. This machine evidently does not constitute a whole with meaning (at least as far as can now be ascertained). Its basic order is that of independently existent parts interacting blindly through forces that they exert on each other. The ultimate implications of this view of universal order are, of course, that man is basically insignificant. What he does then has meaning only insofar as he can give it importance in his own eyes, while the universe as a whole is basically indifferent to his aspirations, goals, moral and aesthetic values, and indeed, to his ultimate fate.

It is clear that these two views will in the long run have very different implications for our general attitude to life which can be profound and far reaching, e.g., man tends to feel much more at home psychologically with an organismic view of the universe than with a mechanistic view. Toward the end of this essay we shall discuss some of these implications in more detail. But for the present we merely call attention to the fact that a mechanistic notion of order has come to permeate most of modern science and technology, and for this reason alone it is now a major factor in human life throughout the world.

2. *Mechanism in Physics*

It is in physics, however, that the mechanistic world view obtained its most complete development, especially during the nineteenth century when its triumph seemed almost complete. From physics, mechanism has since then spread into other sciences and into almost all fields of human endeavor. So some examination of the form that mechanism has taken in physics is called for, if we are to understand what has by now become a more or less dominant world view, which deeply affects all of us. In such an examination, the correctness and necessity of mechanism may be evaluated and criticized, especially with regard to whether or not the actual state of knowledge in physics (and elsewhere) continues to sustain and support such a view, as well as to whether or not alternative views are now possible.

We shall start by listing the principal characteristics of mechanism in physics, and in order to make the meaning of this view clearer, we shall contrast some of its main features with those of an organismic view.

(1) The world is reduced, as far as possible, to a set of basic elements. Typically, these have been taken as particles, such as atoms, electrons, protons, quarks, etc., but to these may be added various kinds of fields that extend continuously through space, e.g., electromagnetic, gravitational, etc.

(ii) These elements are basically *external* to each other, not only in being separate in space, but more important, in the sense that the fundamental nature of each is independent of that of the other. Thus, the elements do not grow organically as parts of a whole, but rather, as indicated earlier, they may be compared to parts of a machine, whose forms are determined externally to the structure of the machine in which they are working.

(iii) As also pointed out earlier, the elements interact mechanically, and are thus related only by influencing each other externally, e.g., by forces of interaction that do not deeply affect their inner natures. In contrast, in an organism the very nature of any part may be profoundly affected by changes in the activities of the other parts, and so the parts are basically *internally* related. Of course, in a mechanistic view the existence of organisms is admitted, but it is assumed, in the way described above that their behavior can eventually be explained as the results of interactions of constituent molecules (such as DNA), which are in turn ultimately reducible to structures of smaller particles (such as electrons, protons, or quarks), that will finally be discovered to be related only externally and mechanically.

In addition, it is admitted that such a goal is yet to be fully achieved, as there is much that is still unknown. But it is essential for the *mechanistic reductionistic program* to assume that there is *nothing* that cannot eventually be treated in this way. Its adherents point to its success thus far to justify this assumption. But of course this is in no sense a proof. And so, to suppose that this assumption holds without limit is basically an *article of faith*, which permeates the motivation of most of the modern scientific enterprise, and thus provides much of the human energy that is needed to carry it out. This is a modern counter-part of earlier faith in religious beliefs, based generally on more organismic types of world view, which also in gave energy to vast social enterprises.

How far can this modern faith in mechanism be justified? Of course, there can be no question that it works in a very important domain, and that it has brought about a revolution in our mode of life. Indeed, as indicated earlier, during the nineteenth century, there seemed to be little reason to doubt this faith, because of what appeared to be several centuries of successful application of the world view. It is therefore hardly surprising that physicists of the time commonly had an unshakable confidence in its correctness. To illustrate this, we may refer to Lord Kelvin, one of the leading theoretical physicists of that period, who expressed the opinion that physics was more or less complete in its development. He therefore advised young men not to go into this field, because further work in it would only be a matter of "refinements in the next decimal points." He did, however, mention two small "clouds" on the horizon. These were the negative results of the Michelson-Morley experiment, and the difficulties in understanding black-body radiation. It must be admitted that Lord Kelvin was at least able to choose his "clouds" properly, for these were precisely the points of departure for the development of relativity and the quantum theory, which together brought about a radical revolution in physics, overturning the entire conceptual structure of the hitherto dominant Newtonian (as classical) physics. This clearly illustrates the danger of complacency with regard to our world views, and makes it evident how necessary it is constantly to have a somewhat provisional, exploratory and inquiring attitude with regard to them.

3. *Relativity as an Important Step Away from Mechanism*

This is not the place for a detailed explanation of how, precisely in the field of physics which had been the central support for a mechanistic world view, there began early in the twentieth century a new development in which this mechanistic view was eventually seen to be completely inadequate.

We shall give here only a brief non-technical sketch of this development.

We begin with the theory of relativity which introduced a number of fundamentally new concepts of space, time and matter. For our purposes, however, the main new idea of Einstein is that he replaced the notion of separate and independent particles as basic constituents of the universe by that of fields that spread continuously through space. We may illustrate these ideas in terms of the analogy of flow of a fluid, such as water. Within such a fluid, there may be a *vortex*, which is a constant recurrent pattern of form of movement of the whole that is stable.

Because the movement gets weaker as one goes further away from the center of the vortex, this pattern does not significantly involve far-away features of the flow, and thus, it has a certain relative independence of what is happening in distant parts of the fluid. The *form* of the movement may therefore be conveniently abstracted in our minds and given a name (i.e. vortex): as if it were a separate entity. But evidently, this is only a way of talking and thinking, and is not a description of what is actually happening (which is an unbroken flow of fluid).

To see in more detail what our analogy means, let us now think of two vortices that are far from each other. Their flow patterns affect each other only weakly, so that they are nearly independent of each other.

Let us now consider bringing the two vortices closer, so that the patterns of movement affect each other more strongly. If they are brought still closer, they may then merge into a single complex vortex structure.

What we have here is an example of unbroken *wholeness in flowing movement*. Separate "entities" (such as vortices) are, in this view, relatively constant and independently behaving forms, abstracted from the whole, in perception and in thought.

All this was, of course, well known to nineteenth-century physicists. However, it was generally implied in the work of most of them that real fluids, such as water, are constituted of myriads of elementary atomic particles, which "flow" only in an approximately continuous way (like grains of sand in an hour glass). The reality underlying the macroscopically observed fluid was thus considered to be a structure of discrete mechanical elements, in the form of particles.

On the basis of the theory of relativity, however, Einstein gave arguments showing that such elementary particles would not be consistent with the laws of physics, as developed in this theory. Instead, he proposed a set of continuous fields pervading all space, in which "particles" would be treated as stable and relatively independent structures of limited regions in which the field was strong.

These would, like vortices in water, gradually shade off into weaker and weaker fields. Such structures of field were shown mathematically to move through space as a stable unit (like a smoke ring vortex). As two of them came closer together they would begin to influence each other more and more. Eventually, they would merge. Each so-called particle is thus explained as an astraction of a relatively independent and stable pattern of movement of fields, spreading out through space, with no breaks anywhere. The universe is in this way seen as an *unbroken whole in flowing movement*.

This approach contradicted in an important manner the assumption of separate "elementary" constituents of the universe that had been characteristic of the mechanistic world view. Yet, in doing so, it still retained some of the essential features of mechanism. For the field elements at different points in space were considered to be separately existent and not internally related in their basic natures. The separate existence of these basic elements was further emphasized by the assumption that they were only *locally* connected (i.e., the field at a given point could be affected only by fields at infinitely close neighboring points). The over-all field was thus viewed as a type of mechanical system that was more subtle than a system of particles. Nevertheless, the field approach was still an important step away from the mechanistic world view, even though it remained within the general framework of this kind of view.

4. *The Overturning of Mechanism in the Quantum Theory*

The quantum theory, however, actually overturned mechanism in a much more thoroughgoing way than did the theory of relativity. We now give here its three main features:

(i) All action, all motion, is in discrete indivisible units, called *quanta*. (Hence the name, quantum theory). For example, in the early forms of the theory due to Bohr, an electron had certain sets of discrete possible orbits. It was assumed that the electron jumps from one of these orbits to another, without continuously crossing the intervening space. Action of every kind is of this discrete indivisible nature (whether of particles or of fields). The apparent continuity that is commonly observed arises because the individual quanta are very small. An ordinarily visible movement therefore has a very large number of discrete jumps, each too small to be perceptible (except possibly with the most sensitive of instruments). So all apparently continuous large-scale (i.e., classical) motions are to be understood as constituted of discrete steps. Such a notion certainly contradicts the older classical concept of continuity of movement, which is at the very basis of the mechanistic ideas of Newtonian physics.

(ii) All matter and energy are found to have what appears to be a dual nature, in the sense that they can manifest either like a continuous wave or like a discrete particle, according to how they are treated in an experiment. For example, the electron, which is classically a particle, can under suitable conditions also behave like a wave, but the wave-length is so small that except in a very refined observation this does not show up. Similarly, light, which is classically a wave, can under suitable conditions behave like a particle (or a collection of particles) but the energy of such particles is so low that except in a very refined observation this also does not show up. The fact that any system can show either wave-like or particle-like characteristics, according to the general condition of its environment (which is in this case the observing apparatus) is clearly not comparable with mechanism. Indeed, the variation of the fundamental nature of an entity according to such conditions is much more like what is encountered with living and even conscious organisms than it is like what is to be expected from a machine.

(iii) One finds a peculiar new property of *non-locality of connection*, i.e., a close rapport between particles or other elements (e.g.,fields) that may be distant from each other. This violates the classical mechanical requirement of *locality of connection* (which we have already mentioned in the discussion of Einstein's notions regarding the nature of the field). This latter requirement is that the *basic* elements constituting the universe (whether particles or fields) are strongly connected, only when they are in contact in space, or else, infinitesimally close together.

To bring out further how these three key features of the quantum theory contradict the basic mechanistic assumptions, let us first consider the fact that all action and interaction is through discrete indivisible quanta. This means that all parts of the universe are connected by indivisible links, so that there is no way ultimately to divide the world into independently existent parts (in principle, this extends even to the observer and what is observed). Moreover, the fundamental nature of each part (wave or particle) depends inextricably on this web of indivisible quantum links that are its context. And finally, since indivisible interconnection may extend even to distant regions of space, it follows that the very nature of each part may depend significantly on what is happening in places that are quite far from it.

Of course, all of this is in general evident only under highly refined modes of observation. At ordinary levels of refinement (including classical or Newtonian physics) familiar mechanistic conceptions will furnish a good approximation. And so, we understand why the mechanistic programme worked fairly well for hundreds of years (i.e., until observations were refined enough to reveal the more fundamental non-mechanistic structure). Yet, if we wish to go deeper, and above all, if we wish to understand the basic nature of the universe, we have to notice that this carries us beyond the limits of what can be done with the mechanistic programme. For the notion of analyzing the world in terms of independently existent elements, whose fundamental natures are *external* to each other, has broken down.

Here, it has to be added that the quantum theory denies another well known feature of classical physics, i.e., its complete determinism. This latter is commonly illustrated by Laplace's idea of a demon who could know the initial positions and velocities of all the particles constituting the universe. Using Newton's laws of motion, such a demon could in principle calculate the behavior of these particles for all time, and thus he could know both the whole past and the whole future of the entire universe.

The laws of the quantum theory would not, however, permit such a calculation, because they are *statistical*. That is to say, they give only probabilities that certain things will happen, but do not determine in detail what actually will happen in each case. And thus, the quantum laws are not deterministic, though in the limit of a structure large enough to be observable by ordinary means, so many discrete steps are involved that the predictions of probability laws become nearly deterministic (as insurance statistics can be used to predict fairly accurately the fraction of people in a large group who will die in a certain way, though it can say nothing about precisely what will happen to each individual).

It has to be emphasized, however, that this question of determinism vs. indeterminism has little or no relationship to that of mechanism vs. non-mechanism. For the essential point of mechanism is to have a set of fundamental elements that are *external to each other* and externally related, in the sense described earlier. Whether these elements then obey deterministic or statistical laws does not affect the question of the mechanical nature of the basic constituents (e.g., a pin-ball machine or roulette wheel that would operate according to "laws of chance" is no less mechanical than is a machine whose behavior is completely knowable and predictable).

4. *Unbroken Wholeness--A Non-Mechanistic View Compatible with Relativity and with Quantum Theory*

We shall now consider the question of how the quantum theory and relativity bear on each other, with regard to the mechanistic world view. This question is not easy to discuss, because it does not seem to be possible to relate the *basic physical concepts* of the two theories in a consistent way. For relativity requires strict *continuity,* strict *determinism,* and strict *locality,* in the formulation of its laws, while quantum theory requires *discontinuity, indeterminism,* and *non-locality* in such formulations. Thus, they appear to be in absolute contradiction. And indeed, within the present general frameworks of these two theories the two sets of *physical concepts* have never yet been brought together consistently in a unified theory.

If we are to approach the question of looking at relativity and the quantum theory together in a coherent way, we may be led to consider a new kind of question. Instead of focusing on how the basic concepts of these two theories contradict each other, let us instead ask what they have in common. What is common to both is unbroken wholeness of the universe. Each has this wholeness in a radically different way. Yet, if wholeness is their common factor, this is perhaps the best place to start in the search for new physical ideas, by which we may understand the novel and subtle features to be seen in these theories (and in the essentially mathematical formulation of their union in quantum field theories).

We have seen thus far that each world view holds inseparably within itself its own basic notions of order. The ancient Greek view incorporated, as pointed out earlier, the order of increasing perfection from Earth to Heavens, and the order implied in the idea that each part is striving to reach its proper place and to fulfill its appropriate function in the whole. However, the world view implicit in Newtonian physics is based on the notion that such an order is totally irrelevant, and that what is important is the mechanical order of successive positions traversed by each particle, and of the strength of forces, which they exert on each other. This latter order is now expressed mathematically in terms of *coordinates*, originally introduced by Descartes. (These are grids, by means of which the locations of points can be accurately specified in terms of numbers.) As the very word indicates, such coordinates are means of describing order and, of course, the order is of just the kind that is needed for thinking about a universe that would be basically mechanical in its nature.

We are in this way led naturally to the question: Is it possible to develop a new order that is suitable for thinking about the basic nature of a universe of unbroken wholeness? This would perhaps be as different from the order of mechanism as the latter is from the ancient Greek notions of an order of increasing perfection in an organismic universe.

This, however, brings us to the further question: What *is* order? We may begin by noting that all that we do presupposes order of *some* kind, so that a general and explicit definition of order is not actually possible. Nevertheless, we may indicate what this notion means with the aid of a number of typical examples. These include the order of the numbers, the order of points on a line, the order of functioning of a machine, the more complex and subtle order of functioning of an organism, the many orders of tones in music, the order of a language, the order in thinking, etc. Evidently, the notion of order covers a vast and unspecifiable range, and we may take it that, as implied above, we already have some sense (largely tacit rather than verbally expressed) of what order is.

5. The Holograph as an Illustration of an Order of Unbroken Wholeness

Since most of this tacit sense of order is ultimately based on perceptual experience (as is evident from the examples considered above), one may ask whether there is not some typical example or analogy in our experience that could serve to indicate the new order of unbroken wholeness that is being suggested here. With regard to the question, it may be pointed out that the operation of scientific instruments has often played a key part in helping to make certain notions of order vivid and clear.

Thus, the lens, for example, is a device that works an image in which each point, P, of an object corresponds (in a high degree of approximation) to a certain point, Q, in the image. When the image is recorded on a photograph, this constitutes a kind of *knowledge* of the object. With the aid of telescopes, microscopes, very fast or very slow cameras, etc., this sort of knowledge through correspondence of points has been extended to things that are too far away, too small, too fast, too slow, etc., to be seen with the naked eye. And thus, people have been led to think that everything could ultimately be known in the form of a correspondence of separate elements. Instruments based on the lens have thus given an enormous impetus to the mechanistic way of thinking, not only in science, but in every phase of life.

Have any instruments been developed that would similarly help point vividly to a way of thinking that is compatible with unbroken wholeness? It turns out that there are several. I shall begin by describing the *holograph* (invented by Denis Gabor). The name is based on two Greek words, "*holo*" meaning "whole," and "*graph*" meaning "to write." Thus, a holograph "writes the whole."

This instrument depends on another device called a *laser*, which produces a beam of light in which the waves are highly ordered and regular (in contrast with those from ordering sources, in which they are rather chaotic). Light from such a laser then falls on a half-silvered mirror, which transmits part of the original beam and allows it to go straight ahead, while another part is reflected onto an object. This latter scatters the waves, and these scattered waves propagate onward until they meet the transmitted part of the original beam. Here, the two sets of waves combine to produce what is called an interference pattern. This is actually a very complex distribution of wave motions, generally too fine to be seen in detail with the naked eye. The interference pattern is then recorded in a photograph. Of course, this does not look like anything in particular, and as indicated above, it is indistinct and hardly visible. The next step is to illuminate the photograph with laser light similar to that which was used originally.

A pattern of waves then emerges from the photograph that is similar to that which comes from the original object. If the eye is placed in front of the photograph, an image appears of a three-dimensional object. But this is not the main point for our purposes here. The main point is that each part of the photograph can yield an image of the *whole object* (as if seen through a window the size of the illuminated region). We can see that waves from the whole object enter each region of the photograph, and so, the pattern that they produce contains information about the entire object. In this way, a recorded holographic image contains a kind of knowledge in which there is no point-to-point correspondence with the object. Our attention is thus called to a new form of knowledge, in which information about a whole is *enfolded* in each part of an image. (To give some preliminary idea of the meaning of the word "enfold" in this context, we can usefully consider how the points of contact made by folds in a sheet of paper may contain the essential relationships in the total pattern displayed when the sheet is unfolded.)

6. *The Enfolded or Implicate Order*

Of course, in the example given above, the photograph is only a static record of the light, which is a movement of waves. The actuality that is directly recorded is this movement itself, in which information about the whole object is *dynamically* enfolded in each part of space, while this information is then *unfolded* in the image.

A similar principle of enfoldment and unfoldment can be seen to run through a very wide range of experience. For example, the light from all parts of a room enfolds information about this whole in each region and an eye placed in any position can then unfold this information into an image on the retina (which is further enfolded and unfolded in a vast range of structures in the brain and nervous system). Similarly, the light entering a telescope enfolds information about the whole universe of space and time. And more generally, movements of waves of all sorts enfold the whole in each part of the universe.

The principle of enfoldment and unfoldment may readily be observed in much more familiar contexts. Thus, for example, the information out of which a television image is formed is enfolded in a radio wave which carries it as a signal. The function of the television set is indeed just to unfold this information for display on the television screen. This was especially evident in older television sets, which had an adjustment for synchronism. When this adjustment was faulty, the image was found to fold up into a nondescript pattern (in some ways similar to the interference pattern in the holograph). A suitable adjustment then allowed the image to unfold in its proper order.

Now, in the mechanistic world view, all these examples of enfoldment and unfoldment are well known. But they are explained by saying that the primary reality is ultimately the basic set of independently existent elements (particles and fields) that make up the universe, while enfoldment and unfoldment is only a secondary aspect of reality. What is being proposed here instead is that the movement of enfolding and unfolding is ultimately the primary reality, and that the objects, entities, forms, etc., which appear in this movement are secondary.

How is this possible? As has already been pointed out, quantum theory shows that all the so-called particles constituting matter in general are also waves similar to those of light (thus, one could in principle make holographs using beams of electrons, protons, sound waves, etc.). The key point is then that the mathematical laws of the quantum theory that apply to these waves and therefore to all matter can be seen to describe just such a movement as has been explained above, in which there is continued enfoldment of the whole into each region, along with unfoldment of each region into the whole. Although this may take many particular forms, some of which are now known, and others of which are not, such movement is as we have seen universal. We shall call this universal activity of enfoldment and unfoldment the *holomovement*. The proposal is then that the holomovement is the basic reality, and that all entities, objects, forms, etc., as ordinarily seen, are relatively stable, independent and autonomous features of the holomovement (as the vortex is such a feature of the flowing movement of a liquid). The basic order of this movement is therefore enfoldment and unfoldment. So we are looking at the universe in terms of a new order, which we shall call the *enfolded order* or the *implicate order* (the word "implicate" is based on a Latin root, *"plicare,"* meaning "to fold," as shown in words such as multiplication (multifolding) and replication (re-folding).

In the implicate order, everything is folded into everything. Here, it is important to note that the whole universe is in principle enfolded into each part *actively* through the holomovement. This means that the dynamic activity, internal and external, which is fundamental to what each part *is,* is based on its enfoldment of the entire universe, and thus of all other parts (of course, the various parts are enfolded in each other in different ways and in different degrees, that are characteristic of each part, but the basic principle of enfoldment of the whole is not thereby denied).

So the enfoldment is thus not merely of a superficial or passive kind. Rather, we emphasize once again that each part is, in a fundamental sense, internally related *in its basic activity* to the whole and to all the other parts. The mechanistic idea of external relationship as *fundamental* is therefore denied, though of course, such relationships are still considered to be real, but of secondary significance. That is to say, the order of the world as a structure of things that are basically external to each other comes out as a secondary order through the activity of unfoldment which emerges from a deeper and more inward implicate order. The order of elements external to each other will then be called the *unfolded order* or the *explicate order*. Thus, the usual way of thinking on the subject is, in effect, turned upside down, and this is how we arrive at the new notion of implicate order.

The holograph is, of course, only a particular example of an implicate order. Its value in the present context is that it provides a good analogy to how the implicate order is relevant to the quantum behaviour of matter. This analogy is particularly good, because, as we have indicated earlier, the laws of propagation of the kinds of waves that are associated with the basic quantum laws are also capable of being compatible with the theory of relativity. And thus, we see that the implicate order is able to have a significant bearing on both of the most fundamental theories of modern physics.

But, of course, analogies are necessarily limited, since by their very nature, they are similar only in some ways to the subject of interest, and are different in other ways. One of the principal limits of the analogy of the holograph (at least as it is usually analyzed) is that it does not adequately take into account *all* of the quantum properties of the waves that are involved. In particular, what it fails to consider is that the energy of these waves is in discrete units or quanta (called photons) and that, the fields at different places do not in general have purely local connections (in the way that we have previously mentioned). The holographic analogy therefore still misses some of the essential features of quantum wholeness. To make a more accurate analogy, one would have to use the modern relativistic quantum field theory, but, of course, such a treatment would (at least at present) be so abstract and mathematical that it would not be of much help as an analogy whose purpose is to make the meaning of the implicate order intuitively and imaginatively clear.

7. *Extension of Implicate Order to Life and Consciousness*

It is possible actually to produce an indefinite number of additional analogies to the implicate order. It is not our purpose, however, to go further into such details here. Rather, what we shall now do is to go on to discuss the more general significance of the implicate order, beyond physics.

First, let us consider the example of a living being, such as a plant grown from a seed. The seed, however, makes a very small contribution to the substance of the fully grown plant, and to the energy needed to make it grow. These contributions come from air, water, soil, and sunlight. According to modern ideas of genetics, the seed has *information*, in the form of DNA, which is transmitted to the matter out of which a plant is eventually formed. We have already been led to use the notion of implicate order for matter in general. It is continually unfolding and folding again into a background, and thus, even inanimate matter (e.g., an electron) is, as it were, constantly "remaking itself." With further information from the seed, however, it unfolds to make a plant instead (which can then make more seeds for new plants).

We now apply such a notion to a system, consisting of a large number of plants, for example, a forest, which we consider covers a long period of time. Let us imagine a visit to such a forest every hundred years. One would see changes, as if trees had "moved," "crossed space" and "transformed." This is also a good image for how electrons and other fundamental particles are maintained and changed, according to the implicate order (of course, the scale of time is immensely shorter). And more generally then, we say that all matter, animate and inanimate, unfolds from a greater whole and folds back again into it, in an endless process of replication of forms that are similar but different. So there is, in this respect, no sharp distinction between living and non-living matter.

Let us now go on to discuss consciousness, which we take to include thought, feeling, desire, will, impulse to act, and an unspecified set of further features, some of which we shall discuss later. The question is then: do we find an implicate order in consciousness?

To answer this question, let us first consider the process of thought. In describing this process, we may refer to thoughts that are implicit. The word "implicit' has the same root as "implicate." This suggests that a given thought may somehow contain in it other thoughts that it *implies*, i.e., enfolds. Such implication may, in some cases be equivalent to *entailment* or *inference*, if it obeys the rules of logic. However, this is only a special case of implication (like that of a regular track in the sequence of ink droplets, discussed earlier). Indeed, one finds that implication may have a much wider range of meanings, going from mere association to a sense that one thing "goes with another" and on to a tacit or unstated ground or reason supporting the thought that is implied. All of these may be regarded as enfolded within the thought in question, and as capable of emerging from it through unfoldment.

Here it may be added that language, which is essential to the communication of thought and to its precise determination, may also be seen as being in an implicate order. After all, the word is only a sign or symbol, of little significance in itself. What is more important than the word alone is evidently its *meaning*. Generally, this is determined properly only in a much larger context. For example, the meaning of a given word may be significantly affected by other sets of words, not only near to it, but even quite far away, in the sequence of discourse. This suggests that the meaning of each word (and indeed, of each combination of words, such as a sentence, a paragraph, etc.) is ultimately enfolded in the whole content that is communicated. Such a notion is suggested even more strongly by the fact that often one can sense that a whole sequence of words seems to flow out of a single momentary intention, without the need for conscious choice of their order, essentially as if they had unfolded from something that was already there in the intention.

A further interesting example suggesting enfoldment is to be seen in the fact that, without the need for a "search" in memory we can generally sense whether a word is in common usage in our language or not. Thus, whereas nouns formed out of verbs (e.g., "alternation") usually have in common usage verbs to correspond with them (e.g., "to alternate"), we know immediately that in certain cases, they do not (e.g., "alteration" does not correspond to an allowable verb "to alterate"). The immediate availability of this knowledge does indeed suggest that the totality of a given language is an undivided whole, from which the various words (and, of course, their potential meanings) all unfold.

We see then that a reasonable case can be made for the proposal that thought and the language forms that express and shape it are in an implicate order. Moreover, one can see that in some sense, thought and language further enfold feelings, and that vice versa, feelings enfold thought and language. (Thus, the thought of danger unfolds into a feeling of fear, which further unfolds into words communicating the feeling, and into thoughts aimed at obtaining security.) Likewise, thoughts and feelings together enfold intentions. These are in turn sharpened into a determinate will, and the urge to do something. Thus intention, will, and urge, unfold into action, which includes more thought, if necessary. So all aspects of the mind show themselves as enfolding each other, and transforming into each other through unfoldment and enfoldment. In this way, we have a view in which the mind is not regarded as broken up dualistically or multiply into independently existent functions or elements.

If one is attentive, one can see a great many further clues showing that the mind operates basically in an implicate order, in the general way described above. Moreover, there is also a great deal of modern scientific evidence indicating that sense perception and physical action are primarily in what is in essence an implicate order, and that our consciousness of the explicate order originates in this deeper and more inward implicate order. Here, I can give two examples based on general and common experience.

First let us consider listening to music. Attention shows that while any one note is being played, several preceding notes are still present, in awareness, as a kind of immediate "afterecho" or "reverberation." This is to be distinguished from *memory*, which is what is *recalled* as *recollected*, from a more permanent repository in which it is stored (e.g., remembering notes a minute apart is not perceived as "music," and most of the music is then lost). One can sense that each note, as it starts to fade and turn into a diminishing sequence of "after-echoes," is, in some way, enfolding into various aspects of consciousness, including emotions, associations of various kinds, impulses to move, etc. It is being suggested here that this may be seen as a kind of enfolded order. That is to say, one can sense the co-presence of "after-echoes" and other derivatives of several notes in different degrees of enfoldment (in a structure similar to the interpenetrating ink droplets enfolded through different numbers of turns). The essential point here to which I should like to draw attention is that it is the simultaneous co-presence of several such notes that is at the origin of the sense of flowing movement of the theme, with preservation of its essential identity (which explains why notes that follow each other only after long intervals generally convey neither a sense of flowing movement, nor of preservation of identity of a theme).

The second example that I would like to use here is that of bicycle riding. In this connection, I first call attention to the fact that in order to remain stably upright, one must turn *into* the direction in which one is falling. Michael Polanyi has in this connection pointed out that a simple calculation based on the laws of physics shows that if the bicycle is ridden properly, its angle of tilt and the angle at which the wheel is turned are related by a certain simple formula. But, of course, any attempt *to follow* this formula would get in the way of actually riding the bicycle (though it might be useful in other activities, such as designing new bicycles). What is of key significance here is that the over-all movement whose result is described (to a high degree of approximation) by the formula is the net outcome of an entirely different level of activity involving muscle, nerves, and brain. This activity is extremely complex and subtle, and evidently, one cannot possibly describe in any explicit way just how one does it, or just exactly what is happening. I would like to propose that this activity may also be regarded as a kind of implicate order, which unfolds into an explicate order of motion of the bicycle as described by the formula. The law of the explicate order thus emerges as an abstraction of what is actually a single feature of a much larger explicate order.

As we gain experience in riding a bicycle, what has been learned is stored up somehow as a kind of skill, which constitutes a sort of knowledge. Polanyi calls this *tacit knowledge,* because its essential nature cannot be put in words. Evidently, tacit knowledge (which I have suggested is also implicit or enfolded) is necessary in every phase of life. Indeed, whenever one has actually to do anything whatsoever, such tacit knowledge clearly must come into play. For example, if one wishes to walk across the room, abstract thought may present a mental image of the projected goal that has to be realized, but how one actually does it is just as non-verbalizable as is how one actually manages to ride the bicycle. And for that matter, what one actually does in order to engage in abstract thinking is, if anything, even more indescribably subtle and non-localizable than is what one does to engage in physical movements.

On the basis of all this, I would then propose for further discussion the notion that both mind and matter are ultimately in implicate orders, and that in all cases, explicate orders emerge as sets of relatively autonomous, distinct and independent objects, entities and forms, which unfold from implicate orders. This means that the way is opened up for a world view, in which mind and matter may consistently be related, without adopting a reductionistic position in which one of these would be nothing but a derivative outcome of the other (e.g., materialism, in which mind is just a function of matter, and idealism, in which matter is just a function of mind). Rather, our proposal is that mind and matter both arise from a common ground that is beyond either, and that is ultimately unknown. The essential point, however, is that because they have the implicate order in common, it is possible for them to have a rationally comprehensible relationship. Moreover, because their mode of arising from the ground is unfoldment, they both enfold this ground, and therefore enfold each other, so that their relationship is basically internal. In this way, we have left the way open to acknowledge whatever differences that may be found between mind and matter, without thereby falling into dualism.

In this connection, it should be pointed out that the question of how mind and matter are related has long been one that has perplexed those who have inquired into it seriously. Descartes has given an especially clear and sharp formulation of the difficulties involved. He considered matter as extended substance (i.e., as existing spread out in space in the form of separate objects). Mind he discussed in terms of thinking substance which is not separated and extended in this way. For, although we are capable of clear and distinct thoughts, they do not exist as separate and extended elements in any kind of space. Descartes felt that the two substances are so different that there is no way to formulate their relationship clearly. The problem of how they are related is to be solved, he proposed, by bringing in God, who created both, and who is thus the ground of their connection (e.g., God puts clear and distinct thoughts into our minds, which may correspond correctly to the separate extended objects in space).

Since the time of Descartes, the idea that problems of this kind can be solved by an appeal to the action of God has more or less been dropped. But it has not generally been noticed by those who go on with Cartesian mind-matter duality that this leaves the problem of how the two are related unsolved.

In effect, the implicate order solves this problem of this Cartesian duality, which has indeed pervaded much of human thinking for ages. For instead of saying that there are two orders--the explicate order of extended structure, and something like an implicate order of thinking, it is now being proposed that there is only one universal order for mind and matter, and evidence (scientific and more general) has been given, indicating that it is consistent to say that this is the implicate order. The implicate order thus clearly provides a possible basis for a different approach to the understanding of the nature of the universe, in which ultimately, all is seen as encompassed in a single whole, that includes mind and matter as essentially related.

II. *On Wholeness and How the Parts Fit In*

It has to be emphasized very strongly at this point, that to have an approach of wholeness, such as that given above, does not mean that we must be able actually to capture this totality of existence within our concepts and knowledge. Rather, it means, firstly, that we understand this totality as an unbroken and seamless whole, in which each relatively independent and autonomous entity, object, form, etc., merges along with the others, in a background of ultimately immeasurable extension and depth of inwardness. And secondly, it means that insofar as the wholeness is comprehended with the aid of the notion of the implicate order, the ultimate internality of relationship has necessarily to be taken as basic. Such a notion is indeed also suggested by an organismic point of view, but here, as we have seen earlier, there is no way to exclude the possibility that organisms have a mechanistic base in their supposed constituent particles.

With regard to this question of wholeness and internality of relationships, it is important to keep in mind that "whole" and "parts" are correlative categories, and that each implies the other. Indeed, it is clear that something can be a part, only if there is, potentially at least, a larger whole of which it *is* a part. This is evidently so as much for mechanical parts (formed by external action) as for organic parts (formed in the context of the activity of the whole).

To understand how the correlation of the whole and parts is treated in the implicate order, let us return to the notion of the holomovement. Within the holomovement, as we have seen, each part emerges as relatively independent, autonomous, and stable, and it does so by virtue of the particular way in which it *actively* enfolds the whole (and therefore all the other parts). Its fundamental qualities and activities, internal and external, which are essential to what it *is*, are thus understood as determined basically in such internal relationship, rather than in isolation and in external interaction. This means, of course, that all parts are *internally* related, through such relationship with the whole.

Such internality of relationship is most directly experienced in consciousness. The content of consciousness of each human being is evidently basically as enfoldment of the totality of existence, physical and mental, internal and external. This enfoldment is *active*, in the sense that it enters in a fundamental way into the activities that are essential to what a human being *is*. Each human being is thus internally related to the totality including nature, and the whole of mankind (and he is therefore also internally related to other human beings). What we are further saying here is that the quantum theory implies that ultimately the relationship of parts and whole for matter in general is to be understood in a similar way (what we are also proposing is a similar relationship of mind and matter). We therefore emphasize once again that the implicate order does not deny the significance of parts, but rather that it treats them in its own way, as relatively stable, independent and autonomous sub-wholes. Wholeness is thus put as primary, while the parts are secondary, in the sense that what they are and what they do can be understood only in the light of the whole.

This may be summarized in the principle:

The wholeness of the whole and the parts.

For the sake of completeness of discussion, we may also consider the opposing principle:

The partiality of the parts and the whole.

Evidently, both principles are, in a certain sense, true. For, to deal properly with all aspects of the question, we have to take into account that while parts arise out of the whole, and are in essential relationship with this whole, they are also in important respects distinct. But more generally, I am suggesting that these two categories are not completely symmetrical in their relationships. That is to say, *ultimately*, the wholeness of the whole and the parts is the major or dominant factor, while their partiality is a minor one. We may thus put the emphasis on wholeness by stating our principle as:

The *ultimate* wholeness of the whole and the parts.

The principle that is denied is then:

The *ultimate* partiality of the parts and the whole.

The latter would put distinction and division in the first place, and would make the being and the activity of the whole depend on the prior independent activity of the parts. Such an attitude toward distinction and division is in essence what is behind mechanism, because it effectively denies that the whole has any independent actuality, and implies that it is merely an abstract way of considering an interacting collection of parts.

So the difference between the approaches of mechanism and wholeness is not in whether whole nor parts are included but in which of these is given primary emphasis and which is taken as secondary (rather, as in a musical composition, the entire meaning depends on which of the themes has a major or dominant role and which is minor or secondary).

9. *Wholeness and Fragmentation*

An important consequence of such an approach of wholeness is that it can help to end the far-reaching and pervasive fragmentation that arises out of the mechanistic and world view. One can obtain a further understanding of the nature of such fragmentation by asking what is the difference in meaning of the words "part" and "fragment." As we have seen, a part (whether mechanical or organic) is intrinsically related to a whole. But this is not so for a fragment. As the Latin root of the word indicates, and as the related word "fragile" shows, to fragment is to break up or smash. Thus, to hit a watch with a hammer would not produce parts, but fragments, that are separated in ways that are not significantly related to the structure of a whole watch. Similarly, to cut up the carcass of an animal, as in a butcher's shop, also produces, not parts of the animal, but also fragments, not significantly related to the structure of a whole animal.

Of course, there are areas in which the production of fragments is relevant and appropriate (e.g., the crushing of stones for the making of concrete). But what we are discussing here is that irrelevant and inappropriate fragmentation which comes about quite generally when we regard the "parts" appearing in our thought as primary and independently existent constituents of all reality (including ourselves). A world view, such as mechanism, in which the whole of existence is thus considered as made up of such "elementary" parts will then give strong support to this fragmentary way of thinking, which in turn expresses itself in further thought that sustains and develops such a world view. As a result of this general approach, man ultimately ceases to give the divisions between things their proper significance (e.g., as useful or convenient ways of thinking, indicative of relative independence or autonomy of these things), and instead, he begins to see and experience *himself* and *his world* as actually made up of nothing but separately and independently existing components. Being guided by this view, man then acts in such a way as to try to break himself and the world up, so that all seems to correspond to his way of thinking. He thus obtains an apparent proof of the correctness of his fragmentary self-world view, not noticing that it is he himself, acting according to his mode of thought, who has brought about the fragmentation that now seems to have an autonomous existence, independent of his will and of his desire.

Fragmentation is thus an attitude which disposes the mind to regard divisions between things as absolute and final, rather than as ways of thinking that have only some relative and limited range of validity and usefulness. It leads to a general tendency to break things up in an irrelevant and inappropriate way, and so, it is evidently inherently destructive. For example, though all parts of mankind are now actually fundamentally interdependent and inter-related, the primary and overriding kind of significance generally given to the widespread and pervasive distinctions between people (family, profession, nation, race, ideology, etc.) is preventing human beings from working together for the common good, and indeed, even for survival. When man thinks of himself in this fragmentary way, he will inevitably tend to put his own separate Ego first, or else his own group. He cannot seriously think of himself as internally related to the whole of mankind and therefore to all other people.

Even if he does try to put mankind first, he will tend to think of nature as something separate, to be exploited to satisfy whatever desires people may happen to have at the moment. Similarly, he will think body and mind are independent actualities, and he will go on in his thinking to divide these further, into various parts and functions, each to be treated separately. Physically, this is not conducive to over-all *health* (whose root meaning is "wholeness"). And mentally, it is not conducive to *sanity* (which has basically a very similar meaning), as is indeed shown by an ever-growing tendency to the break-up of the psyche, as neurosis, psychosis, etc.

To sum up then, fragmentary thinking is giving rise to a reality that is constantly breaking up into disorderly, disharmonious, and destructive partial activities. It seems reasonable then seriously to explore the suggestion that a mode of thinking that starts instead from the most encompassing possible whole, and goes down to parts (sub-wholes) in a way appropriate to the actual nature of things, would tend to bring about a different reality, one that was orderly, harmonious, and creative. But for this actually to happen, it is not enough that we explore this notion only intellectually. It must also enter deeply into our intentions, actions, and indeed, into our whole being. That is to say, we have to mean it, with all that we think, feel, and do. To bring this about requires an action going far beyond what we have discussed here.

Birkbeck College
University of London

BOHM'S CONTRIBUTION TO FAITH IN OUR TIME

John B. Cobb, Jr.

At a conference at Bellagio, Italy, about ten years ago, I was privileged to take part with a number of leading biologists and philosophers in a discussion of evolution. There were many fine papers and the discussion was often rich. But there was one presentation which was, for me and I think for others, the high point of the conference. That was the lecture by a physicist, David Bohm. We sensed very quickly that we were in the presence of a true thinker who was probing behind the level at which most of our discussion, and almost all of contemporary science, operated.

It was not easy to understand what he said. But as it began to make sense to me, I had the joy of a certain kind of recognition. The language and imagery were quite new to me, and I felt the keen disadvantage of knowing so little physics, but the vision that emerged had marked affinities with the one I had learned from Alfred North Whitehead. I saw that Bohm had come to his conclusions quite independently of Whitehead. All the better. He had pressed in the same direction as Whitehead toward a more fundamental view, he had asked some of the same questions. His answers had deep affinities, despite the intervening half century of explosive development in the science from which they both worked. This suggested that these developments did not, as some had argued, invalidate Whitehead's realism and his basic conceptuality. There might be a chance to test it, develop it, and bring it up to date.

I do not mean that I expected Bohm to engage in that task self-consciously. He was purusing his own questions in his own way. That was as it should be. But because the questions were so similar, the results could function for a Whiteheadian both as reassurance and as correction. It also seemed that there were points at which Whitehead's conceptuality might aid Bohm in the development of his own ideas. I still think this may be the case. And in this paper I want both to explain something of the agreement and appreciation that I feel, and to raise the question as to whether there are not still some points on which Bohm could express his vision more clearly and adequately.

I have indicated my indebtedness to Whitehead. Most of what I will say expresses this indebtedness. However, I will talk directly about my own convictions and concerns rather than expounding Whitehead.

Bohm's deepest dissatisfaction with the dominant form of quantum theory is that it remains at the phenomenal level. He wants to know not only how to predict what the appearances will be under certain circumstances but also what is going on whether observed or not. He has succeeded in giving a realistic account of quanta which accounts for all observations. He is ignored because his account does not immediately suggest additional predictions which can be tested. A scientific community that has lost interest in reality regards the realistic theory as of no advantage in itself. To generate the further theories that the realistic approach is capable of producing would require a community of scientists working together with the new model. But the failure to appreciate the advantages of a realistic approach means that no such community emerges. Bohm is left to work largely alone.

Bohm notes that the situation is accidental. If the realistic theory had preceded the phenomenal one, the course of scientific development might have been quite different. Scientists not long ago were open to either type of theory. But at least in quantum theory the phenomenal stance won out because the theory was first formulated successfully that way.

He notes also that there is a general bias in that direction because of the influence of empiricism and positivism. This may be even more decisive than he indicates. At least in much of the wider culture Hume and Kant together, for philosophy and for those fields most influenced by philosophy, virtually suppressed the realistic inquiry into what is.

The relations here are circular. Philosophical positivism about the natural world encouraged a positivistic science. The success of positivistic science provides prestige for the positivist stance in general.

I have begun with this point because I am convinced that its importance for Christian theology is enormous. This may be more apparent to Protestants than to Catholics, since Catholic theology has remained, so far as I can see, much more consistently realistic. Kant has had his effect on Catholic theology too--I think especially of transcendental Thomism--but the God of transcendental Thomism retains being in a way that is not always clear among Protestants who have tried to adjust themselves to the "critical" philosophy. It took the strong reaction of Karl Barth to restore the real God to centrality, and as Barth's influence wanes, once again God is spoken of chiefly in immediate relationship to human experience, not as One Who Is regardless of human existence.

More obvious than the effect of phenomenalism on belief in God is the effect on ethics with respect to the natural world. Most phenomenalists exempt other human beings from the general limitation of all things to phenomena; so it makes sense, following Kant, to speak of our obligations one to another. But since nothing but the human is real, there can hardly be any question of an ethical relation to other creatures. Phenomenalism has excluded consideration of the right treatment of animals and the biosphere as a whole even more thoroughly than did Cartesian dualism.

I agree with Bohm not only that the victory of phenomenalism is a matter of historical accident, but also that phenomenalism perpetuates itself more by indoctrination than by argument. No one can really live as a phenomenalist. I recall Carnap, when I studied with him at the University of Chicago, distinguishing the way he thought as philosopher and in daily life. This did not seem to bother him. But I find the idea that one's philosophy need not account for one's real beliefs preposterous. This idea has certainly contributed to the irrationalism of our culture, of our politics, and of our religion. I have never met anyone who can answer, without irrational arbitrariness, the questions that a critic of phenomenalism finds it necessary to ask. It almost always becomes clear that phenomenalism is a dogma employed in a restricted area to avoid having to deal with hard questions.

No doubt I am being harsher than Bohm, but my intention is simply to agree with him and to support his further inquiries as important not only to physics but to human life and faith. As Bohm indicates, we humans become to a large extent what we enfold of the world, as we believe it to be. To live in a world of phenomena is profoundly to impoverish and delude ourselves. The success of Bohm in showing that the quantum phenomena can be interpreted realistically, that the continuing quest for an intelligible cosmology is justified, is by itself an enormous contribution.

If we agree that we live in a real world and that we should try to understand it as it is, then we confront the next question. What is the world really like? For two centuries the answer was that it was a mechanical world of matter in motion. Indeed, it was the breakdown of that worldview that led to phenomenalism. Kant was sure that science had to think of the world that way, and he could justify that vision only if its basic principles were written into the nature of experience and thought as such. To this day philosophers and scientists have difficulty coming up with ideas about the real world that break free from mechanistic images. Bohm's account of his own intellectual development indicates it was not easy for him. His early pursuit of a causal account seems to have continued mechanistic notions of causality as external determination. Even in his more recent writings he uses the language of determinism that belongs to that worldview, even though he insists he does not mean mechanical determinism.

I believe that there is some confusion in his writings between determinateness and determinism. Any realism must hold that reality has a determinate character. The uncertainty principle cannot mean that reality is not what it is. Further there is a broad sense of determination in which one must also say that finally each thing has been determined to be just what it is. This leads Bohm, as I read him, to speak as if he is a determinist.

But determinism is the doctrine that a state of affairs is determined to take the form it does by *antecedent* conditions. The alternative for the realist is to hold that there is an element of self-determination such that the outcome of a set of conditions is indeterminate until the outcome determines itself. Now self-determination is freedom, and if I understand Bohm correctly he intends to affirm this. To do so will require a clearer rejection of temporal reversibility and of the determinateness of the future than Bohm has yet provided. But overall Bohm's vision would be strengthened in this way. The fact that there is self-determination explains why, from the point of view of those who seek the full explanation in antecedent conditions, there is always indeterminism in the sense that the exact outcome, while determinate, only becomes determinate when it occurs. It is not pre-determined.

However this may be, Bohm breaks unequivocally with the mechanistic worldview by his assertion of internal relations. The entities of which a mechanism is made up must be conceived as matter, and units of matter can relate to one another only externally. They can push and pull. That which is capable of internal relations, that is, of being affected in its constitution by others, is not mere matter. It is more like organism.

This nonmaterial understanding of the real world is expressed clearly also in Bohm's idea of non-local causality. If one conceives of the world as material, one can have no place for action at a distance. Einstein's understanding of space-time has a quasi-material character, and accordingly he was not sympathetic with this feature of Bohm's theory despite his shared commitment to realism. But once one has truly rejected the idea that the world is made up of material particles, then the question of how its units are related to one another is open to investigation with no predetermined assumptions that all influence between distant entities must be mediated by those that are spatially between them.

The introduction of the idea of internal relations transforms the relation of whole and parts. A machine is, of course, more than the sum of its parts, in that the organization of all the parts in that particular way enables it to perform functions that are not simply the sum of the functions of the parts. But the parts are little affected by their role in the machine. A particular bolt can be removed from one machine and placed in another, and we do not think of it as significantly altered thereby. It is only externally related to the other parts of the machines.

If, however, an entity is constituted by its relations to others and to the whole society of which it is a part, then it cannot be removed to some other context without ceasing to be, fundamentally, what it is. Bohm says that the parts are organized by the whole.

While I agree fully with Bohm at this point, I think he overstates his case a little. Reacting against the idea that the parts together constitute the whole, he overstates the priority of the whole. It seems to me more accurate to hold the balance. Whole and part are related in a polar way. Every change in the whole changes the parts, and every change in any of the parts changes the whole. Neither exists apart from the other.

Similarly, I am not clear why the way in which the parts are internally related to one another must be weaker than the way they are internally related to the whole. Indeed, I wonder whether much that Bohm attributes to the relation to the whole is not a product of the relations to all the parts that jointly constitute that whole. I realize that I am asking this without understanding the physics and mathematics that lead Bohm to speak in this way. But I am deeply convinced that we live in world of interconnected entities in a very strong sense, and I worry about moving too quickly to the interrelations between whole and part without giving due weight to the interrelations among the parts. Religiously and ethically this can lead to a mysticism that pays too little attention to the concreteness of history and of the neighbor.

Bohm sees that there are analogies between the processes he describes in quantum theory and those in human experience. Surely he is correct. I only wish he would carry this point through more consistently. If we are realists, and if both our experience and the quanta are part of the reality, then either the reality is deeply divided within itself or else both the quanta and our experience are examples of what the one reality is. As long as the quanta are interpreted mechanically or deterministically, we must opt for dualism to save human freedom. But when the mechanical and materialistic model turns out to be inappropriate for the quanta, the reasons for dualism are undercut. In view of the enormous problems associated with dualism, this should surely be cause for rejoicing.

To affirm that human experience and quanta embody the same reality is not at all to say that there is nothing in human experience that is lacking in the quanta. Human experience is conscious, and there is no reason to attribute consciousness to quanta. Many other differences can be noted. But none of them require us to be dualistic. A human experience is one illustration of the enfolding and unfolding which constitutes all process. A quantum event is another. I would like to suggest that there are dangers in continuing to speak of human experience as mind and quanta as matter, and that the suggestions that mind and matter are related only in the ultimate ground illustrate these dangers. Bohm would do better to abandon this language along with that of determinism. He seems at times to accept the idea that thought affects the bodily condition and is in turn affected thereby, that each enfolds the other. But I am troubled by a lingering preference for relating them through a deeper order they both unfold.

I have another suggestion which seems to me consonant with most of what Bohm says. He writes frequently of process and movement. There seems to be a tension between language that describes this movement as strictly continuous and that which speaks of a constant succession of enfoldings and unfoldings. It seems that the latter is the more fundamental formulation, and that a language that holds firmly to this will prove analytically more powerful. Reality is made up then, not of things that enfold and unfold successively, but of myriads of occurrences of enfoldment and unfoldment. These occurrences, these acts of being, these events are the real units of the real world.

If this can be said, then the remaining ambiguities of Bohm's language about time can also be overcome. He sometimes writes as if time does not apply to the implicate order. This is connected, I think, with the idea that internal relations are primarily with the whole. But is it not better to think of the implicate order first as the enfoldment of all the events that have ever been along with the enfoldment of what is eternal in the whole? In that case, what is enfolded in any two instances will not be quite the same. It will also be clear that what is unfolded in one instance will be enfolded into those that follow. The reality of temporal sequence will be clarified, along with the possibility of real self determination or freedom. The similarities of quanta and human experience will also become more apparent.

It is, I think, clear that my critical concerns are of a piece. The reality of freedom and time and the enfoldment of particular entities by one another all go together. There was an earlier stage of Bohm's thought from which these were all absent, and aspects of that earlier vision continue to intrude on his newer one. At that earlier stage he thought that the solution of the paradoxes encountered in modern physics was to be found at a "deeper" level. That meant that there were "hidden variables," a sub-quantum reality that determined the quantum phenomena. Bohm did not think of this as a return to a mechanical view. The causality he sought was from one level to the other rather than between entities at one level. As quantum phenomena provide a deeper explanation of the sensible world, so there are other levels that provide a deeper explanation of the quantum world. This mode of determination is certainly very different from that of mechanism. Bohm explained it in terms of the relation of an implicate order to an explicate one. But he did see it as quite deterministic. He suggested that if this is not to lead to another form of strict determinism, we might have to think of an infinity of levels.

I confess that I do not find this side of Bohm's speculation attractive. Although I am in poor position to judge, I do not think it has been fruitful in his physical theories. He has recognized that the language of "hidden variables" is misleading, and his actual explanatory proposals seem to make use of particles and quantum potentials rather than deeper levels. The general image of an implicate order, on the other hand, continues to be a fruitful way of looking at many relationships. Most important is its correlation with enfolding as this in turn is clarified as the internality of fundamental relations.

It is on enfolding or internal relations that the newer and now dominant aspect of Bohm's vision focuses, and it is here that his thought is most promising for physics and for religion. It is this emphasis that is primary in his paper at the outset of this volume on implicate order. Internal relations can indeed occur among entities that are in some sense at different levels, but the clearest examples are found at a single level. For example, my experience now enfolds my immediately past experience. It also enfolds distant events. The evidence now is that this enfoldment is not limited to the events in its light cone. It seems that information can be transmitted instantaneously.

The events that make up the life-history of an electron also enfold the events in their light cone and receive instantaneously transmitted information. There is no reason to think that in order for electronic events to be enfolded into events of human experience some third entity must be introduced. We do not need the "deeper" ground. There is undivided wholeness among so-called mental and so-called physical events.

I believe Bohm has said all this. My complaint is that the other imagery of successive layers still intrudes. It sometimes sounds as though the whole were at a deeper level than the quantum and human events and that it is this greater depth that gives it its priority.

I believe that Bohm's physics and metaphysics can be strengthened as he holds steadfastly to two central points: (1) that fundamental relations are internal or enfoldings; and (2) that so-called particles and so-called waves and quantum potentials are both features of the undivided wholeness of a field of events each of which is an instance of enfolding and unfolding. If he does so, he will continue to speak of the whole or holomovement much as he does now, but he will not need an infinity of levels. In my view there can be no level more fundamental than events in their undivided wholeness.

This is not to take a stand on the question of whether there are subquantum events. I have no opinion on the subject of quarks, for example. That is a matter of fact to be determined by evidence. But I doubt that there is need to posit an ether in order to establish an ultimate simultaneity as Bohm has done in a recent paper. Instantaneous enfoldment of information not requiring local transmission on the part of the events making up the field of events suffices.

I have complained that Bohm has so emphasized the internal relations of entities to the whole that he has underestimated their enfolding of one another. In the process I have in my turn overemphasized the reciprocity of relations of creaturely events and the whole. There must, indeed, be some respects in which there is a clear priority on the side of the whole. For example, the whole in some aspects is eternal, whereas no event is in any aspect eternal. Every event has the whole given for it, whereas the whole existed from eternity without that event. The whole is necessary; the particular event is contingent. We could go on. My quarrel is not with recognizing this priority. The whole gives rise to the very existence of the event. This is not a reciprocal relation.

But if, as Bohm and I agree, the whole enfolds the events, then the whole that has enfolded any event is not simply identical with the whole that had not yet done so. We do not have a static whole. We have, in Bohm's language, a holomovement, and that holomovement

46

cannot be abstracted from the events that constitute themselves through their mutual enfoldment and their enfoldment of the holomovement, for the holomovement enfolds them all. Otherwise it is not the holomovement. Here I am appealing for a fuller reciprocity which recognizes that concretely the whole is in its turn constituted by fully real events as well as by eternal aspects.

It is obvious that the modern world has progressively reduced concern for religious thinking and practice. Some have argued that the advance of science has been responsible for this. I believe, instead, that it has been a question of worldviews. Ironically some of the early support for the mechanistic worldview came from those who were afraid that the organic worldview attributed too much to the creatures and not enough to the will of God. They preferred a great machine presided over by its external creator. But this worldview undercut the idea and experience of divine grace, of God actually working redemptively within the world. Further, the hypothesis that the machine was created by something outside itself became less compelling as time passed. We were left with a machine and no God. As the mechanical model in turn became less and less credible, we turned to a positivism which gave us nothing but the phenomena. Any argument from this to a non-phenomenal cause was rejected *a priori*.

Today there are more and more protests against the collapse of all worldview. People can not live by phenomena alone. In this vacuum many are left open to the appeal of very naive and irresponsible ideas. But others are seeking seriously and thoughtfully for a new way of understanding their world.

When people struggle for a new vision of the whole, they are inevitably religious. Whether or not we wish to define religion as the uniting of things, there can be no doubt that it is closely related to that. Whereas secularity fragments, religious thinking seeks to see the whole. To think about how the world really is, is to raise again the questions to which the religious traditions have addressed themselves.

This provides a new opportunity for Christians. When these questions are raised we have a new chance to witness to the Holy One of Israel we have known through Jesus Christ. But there is danger that we will not understand our new opportunity. We will suppose that it is an occasion to trot out old answers to questions developed in relation to different worldviews. These may indeed prove useful and relevant. But if we begin with them rather than with the issues as they arise out of today's world, most of those who quest will only be confirmed in their suspicion that our commitments are to a world that is no more.

When a thoughtful and religiously sensitive physicist such as Bohm shares with us his vision, growing out of meditation on the present state of physics, we have much for which to be grateful. Although, for reasons I have mentioned, his ideas have not caught on in the community of physicists, they have wide resonance among those who are seeking realistic and authentic vision in our time. Our question should not be whether what he says in the end about God fits with our preconceived ideas but whether we can work with him to perfect his intuitions in ways that will also perfect ours.

Bohm has questioned whether the ground of the holomovement to which his vision leads is the personal God. His understanding of the meaning of that phrase is significant, since it reflects very widespread usage, although I doubt that any of the theologians here would use the term in that way. For Bohm, a personal God would be an anthropomorphic God. I trust that we will all agree with Bohm in rejecting anthropormorphism. When we say that God is a person, or three persons, we are not intending to say that God is like a human being or like three human beings. But we must acknowledge that our traditional language has encouraged this type of thinking. Most of us, I think, would say that if, as Bohm believes, the ultimate ground enfolds a supreme intelligence permeated by compassion and love, then God is personal in the sense that is important to us.

But the implications of Bohm's vision go even further. The whole, he tells us, is enfolded to some degree into every part. That is, the whole is within us in important ways. Indeed, I have argued that in his account of events Bohm goes too far in giving the primacy to the internal relation to the whole over the internal relation to other parts. Now it is true that Bohm leaves us with some questions about the relation of the whole to the ultimate ground. But before exploring that complexity, let us note that Bohm's "whole" shares in much of what we think of as divinity and that our relation to this whole is very intimate, very personal, indeed.

Finally, Bohm argues that whereas the whole is in us in a very fragmentary way, we are quite fully in it. In this whole we live, and move, and have our being. The whole to which we respond is also fully responsive to us. Again this is a movingly personal vision.

In our day, and in the spirit of Bohm's own writings, it is a mistake to deal with the ultimate mysteries of the divine in relation to our own Western tradition alone. These mysteries are not revealed to us in any definitive sense in the scriptures. Theologians have dealt with them in the past especially in relation to the philosophies of Plato, Aristotle, and Plotinus as they can be utilized in connection with Biblical thought. Today we need to pay as much attention to the great thinkers and mystics of India as to those of Greece.

I find it striking that in Indian traditions there is a distinction that can also be found, muted, in the West. It is the distinction between the utterly unutterable ultimate and the personal God. In Hinduism it is the distinction between Brahma and Isvara. In Buddhism it is the distinction between the Dharmakaya and the Sambhogakaya. In Christianity it is sometimes the distinction between the Father and the Word, sometimes that between the Godhead and the triune personal God. The *ipsum esse* of St. Thomas corresponds with the former, whereas *ipsum esse subsistens* suggests the latter.

Where this distinction is clearly made, there is a strong tendency to regard the utterly unutterable ultimate, let us call it the impersonal absolute, to be "above" or "beyond" the personal God. For some leaders of these traditions, the latter seems to be little more than a concession to the weakness of believers. But this is by no means the collective witness of these traditions. Far more Indians, Chinese, Koreans, and Japanese have worshipped the personal god than have sought to realize the impersonal absolute. And these have included fine thinkers who have provided theoretical justification for their stance. In the West, where the worship of God has been primary, and the mystical quest for union with the Godhead has been peripheral, it has been the latter that has been on the defensive. In principle there seems to me no reason to subordinate one to the other.

There seems to be in David Bohm's vision also a need for both of these elements. In a general way what he calls the ultimate gound corresponds with the impersonal absolute, whereas what he calls the whole resembles the personal God. If this is so, then perhaps intelligence and compassion should be relocated in the whole, with the ground understood to be beyond all distinctions except as it expresses itself in and is qualified by the whole of which it is the ground. In any case this seems to be a way in which the testimony of the great saints and mystics can be unified with Bohm's vision of physical reality. I am persuaded that a convincing vision can be formed, and that within the vision the Christian testimony can be heard anew and with greater power.

Claremont Graduate School
Claremont, California

MAN AND THE MEANING OF THE WHOLE

Frederick J. Crosson

It is well to remind ourselves at the start that, whatever the vicissitudes of the relationship of science and religion, it is in the Western tradition of religiousness that science developed. It emerged in the space opened up by philosophy which, so to speak, displaced the gods from their active role in the world by the discovery of natures. Cicero's *De Natura Deorum* marks the compromise reached by the reflective man of the classical world for whom the concept of nature had made any explanatory role of the gods unnecessary, but for whom the practical place of the gods in human life was indispensable.

But that space opened up by philosophy--the autonomous character of nature's regularities--corresponded exactly to the incommensurable distance between the infinite and eternal God of Christianity and the created whole. The slow, centuries-long, clear articulation of that culminated in the Middle Ages and provided the stage for the great journey of Dante's *Comedia*. Natural science had its place and role within Christian religiousness for subsequent centuries until each came to misconceive itself and the other.

My object in recalling this history is to remind ourselves that religion and science have passed through a series of relationships with respect to their distinctive teachings: now compatible, now convergent, now conflicting or opposed. The dialetic of the relationships has accompanied, has sometimes brought about, changes in the mode of self-understanding of each of them. So long as the conception of the unity of truth is common to them, so long will that dialectic continue.

We seem largely to have passed through the most recent phase of conflict and to be bordering a period of convergence, for which the most cited sign-post is current theories about the origin of the universe. While I am pleased at the nice things being said on both sides (except for the anachronistic creationist controversy), I confess to a firm desire to remain at the distance of compatibility rather than the embrace of convergence.

Not the deepest reason for this but certainly a sobering one is that mentioned by Professor Bohm, namely that scientific world views have succeeded each other in rather revolutionary fashion and been transient rather than permanent possessions. It would be a hardy soul indeed who would assert that with the "Big Bang" or with quantum mechanics, we have at last broken through to the fundamental level of reality and laid a permanent claim to its understanding.

That is the first *caveat*-reminder, which I want to note by way of introduction. The second concerns not the relations of science and religion but of science and philosophy, specifically of the relation between the properties of micro- and marco-entities, or of macro- and telescopic-entities. The first and third of these may require quite disparate properties to be ascribed in their respective descriptions of reality (e.g. discontinuity and continuity). But it is not clear that these properties must or can be ascribed to the macro- or middle range objects of everyday. Suppose that the statistical laws of quantum mechanics make sense of experimental data in particle physics, so that we can only speak of the probability of an electron being found at a particular orbital location. Does it follow that we should properly speak of the high probability of finding a chair where it seems to be? And if not, must we not ask the same questions about transferring talk of the properties of fields and photons to environments and animals? I find myself, at any rate, puzzled by a number of aspects of the conception of reality as an implicate order, which Professor Bohm has been working out, and I would like to spend the first part of this paper in identifying some issues which seem to need clarification.

I

First, there are unclarities in the basic concepts of enfolding and unfolding (is it always information which is enfolded?).

The holograph is cited in Professor Bohm's essay as exhibiting a new kind of order--new with respect to optical images--in which information about a whole is enfolded in each part of an image. In contrast, in an optical image (e.g. imprinted on a photograph) each point of the image stands in a one-to-one correspondence with the point-parts of the object. To this extent the order exhibited by the optical image is a *mechanical* order, one of disparate parts externally related to form a whole. But the order of the holograph is implicate, because information about the whole in contained (enfolded) in each part of the holographic image.

There appear to be at least three distinguishable meanings of the couple "enfoldment-unfoldment" which do not imply each other, and which are exemplified in different examples given in the essay of Professor Bohm.

(Part of the problem here may be that these are all analogies and not examples, but it is not clear to me that any examples of the implicate order apart from quantum wholeness are offered.)

First, the couple sometimes seems to mean a characteristic of certain wholes in which information about the whole is contained in each of the parts, so that the whole can be "unfolded", recovered from each of the parts. The holograph is an example of this, but not the meaning of a word in a sentence, because here the meaning of the part (word) is enfolded in the whole content.

Second, the couple sometimes seems to mean that what is explicit in the whole is implicit or tacit in the part. Thus what is derivable from a thought is implied by it, whether derived by entailment, association or consequence of meaning, and is described as implicit in the thought. But the structure of the DNA molecule is explicit (we can build a three-dimensional model of it) and the fact that a whole organism can be derived from it in the presence of appropriate materials does not suggest (to me) that it enfolds *tacitly* the information about the whole any more than the binary bits on a floppy disk program implicitly enfold the completed and displayed program.

Third, the couple sometimes seems to mean that what is unfolded is a transient and non-independent part of a whole in which it is embedded and from which it only seems to emerge as a separate entity. So, bicycle riding as an activity described by Polanyi's formula is in reality not isolable from an entirely different level of activity of muscles and nervous system which sustains in existence the overt motion of riding. But the parts of a holograph do not embed the

53

object of which they are said to enfold the information, nor do the wave fronts in each part of a room embed the whole room of which they enfold information.

None of these three senses of the "holomovement" of enfolding and unfolding would seem to be a necessary condition, since at least one example-analogy is a counterexample for each. Moreover, "information" as what is enfolded/unfolded seems sometimes used in its technical sense (the entropy of a message-source) and sometimes in its ordinary semantical sense (e.g. the meaning of a word in a sentence). With respect to the former, it may be noted with Norbert Wiener (in his book *Cybernetics*) that "information is not matter and it is not energy". If information is real in some sense, it is not spatio-temporal.

Perhaps all of these difficulties are related to the fact that analogies always break down at some point. But then we are back with the earlier question: are there any *instances* of the implicate order beyond that of quantum wholeness?

The second topic in Professor Bohm's theory about which I would like to raise a question is that of mechanism, or more specifically, mechanical order.

In contrast to the implicate order, mechanical order is conceived as one in which parts or elements are only externally related to each other, in which wholes are simply the sum of their parts and in which therefore existence or reality is properly ascribed to the elements and not to their configurations. This view of the world order arose already in the ancient world with the atomism of Democritus and Lucretius, and was re-born with the Cartesian mechanical world of the *res extensa*. (Unlike the atomists, however, Descartes' mechanistic universe did not entail a materialist ontology, since the *res cogitans* and God were unextended entities.)

It is worth noting that the world of beast-machines which Descartes proposes involves him in a fundamental inconsistency. For on the one hand he rejects any teleological explanation of natural objects, while on the other hand it is impossible (as Polanyi has noted) to characterize a machine without reference to its function. But a function cannot be adequately specified by giving the physical topography of the machine. Machines belong to the class of objects which can be said to be adequate or deficient, successful or not successful in achieving their function. They require a two-levelled explicans, for while the laws governing the elements may explain why it fails to function properly, they cannot define its successful functioning. We shall return to this issue later in the context of discussing tacit knowledge.

The opposition, not to say dichotomy, of mechanical and implicate order employed by Professor Bohm leads him to speak of the entities of the everyday world as categoreally the same as photons and electrons, i.e. only partially independent of the milieu in which they live and move and have their being. Hence, in deprecating the fragmentation of knowledge and of the human community, he is led to speak of the "parts of mankind" as "actually fundamentally interdependent and interrelated". Mankind forms a whole, a quasi-substantial unity, and indeed even this whole is ultimately only a part (a transient part?) of a yet more encompassing whole: reality itself.

For the moment I shall defer criticism of this claim directly. I want here only to contest the exclusiveness of the division into mechanical and implicate order. Suppose we rejected the conception of purely mechanical order in favor of an Aristotelian ontology of independently existing substances. Are we committed thereby ineluctably to a fragmentation of knowledge and of the human community? It seems not, on either the practical or theoretical level.

Practially, we can establish a unity of order based on the community of nature which all human beings possess and by which they are intrinsically, innately oriented toward a common good. In practice, i.e. in achieving such a human community, this conception seems to provide a no less potentially powerful basis than the conception of all men being "actually fundamentally interdependent and interrelated" as non-independent parts of a single whole. Indeed, insofar as the former conception seems to acknowledge more profoundly and to solicit the freedom of individuals in pursuing a unity, it appears more realistic than the latter. Both conceptions appeal to a unity which is to be acknowledged, not artificially made or chosen, but the former conceives the foundational unity as adumbrating a goal to be implemented rather than a fact to be seen.

Speculatively, we can perceive a deeper unity in the participation of all actual entities in existence or being, and this perception draws closer to the view of all entities as forming, under this aspect, a whole: the created universe. But even this aspect does not join them in substantial unity, for each thing possesses its own existence, its own life, its own understanding. Transient we are, in so existing, but it is at least an open question whether all that we are is resolved back into the dust from which our bodily life emerged.

II

I want now to return to the issue of the implicate order and the relation between our tacit knowledge and the entities which we discriminate by employing it.

It has become a common premise of contemporary philosophy that the meaningful structures of which we are conscious and which come to expression in our utterances are rooted in tacit comprehension of patterns. Those patterns are disclosed in what Wittgenstein called our form of life, and what Heidegger called pre-understanding. It is within a form of life that language is embedded, so that what may seem to exist and have meaning by itself in reality is only the emergent face of an experiential meaning which is not and cannot be rendered fully explicit. The relation of verbal meaning to sentential context iterates this structure of apparent autonomy and real embeddedness.

Professor Bohm is surely right to contrast this understanding with that of Descartes, for whom (as for his successors) ideas can have clear and distinct self-contained meaning, and can correspond to essentially isolable h   s   s   f   ateria   e ts.   In Bohm's view, just as there are no ultimate independent elements in the world (whether animals or sub-atomic particles) so there is no understanding which is not supported by an implicit context.

This symmetry leads to the bold proposal the "mind and matter both arise from a common ground that is beyond either, and that is ultimately unknown". Moreover, since they both enfold this ground they enfold each other and so their relationship is basically internal.

David Schindler has raised some metaphysical questions about the nature of the parts in such a unified field theory of reality, and I share some of his difficulties. What I intend to do here, however, is to go back a stage and think about the symmetry which leads to the proposal, or at least to the thought, of the unity of mind and matter.

It was the Gestalt psychologists, influenced by phenomenology and criticizing the empiricist atomizing of perception, who developed the categories of figure and ground, qualitatively different regious of the perceptual field. And it was Michael Polanyi who took up and generalized that structure as the foundation-stone for an understanding of science.

Let me remind you of some of his extensions of the figure-ground structure and then of some instances of his application of these ideas. (I proceed logically, not historically.)

First of all, the background of simple *Gestalten* is not noticed directly but nevertheless makes an essential contribution to the meaning of the figure which is attended to. The ground can change the elementary qualities of the perceived figure: its color, shape, sound, and the like, etc.

Second, the ground is whatever functions in this way, i.e. as being unnoticed for itself, but radically affecting the perception of the figure. The ground need not be outside the figure. For example, in perceiving the expression on someone's face--pleasure, curiousity, puzzlement--the eyes, mouth, brow, etc. *function* as ground from which the expression emerges.

So let's change our terminology to replace the spatial categories of figure and background. Speak instead of focal and marginal: in the last example, the facial expression is focal, the parts of the face marginal i.e. *marginal to attention*. We are aware of the parts of the face, but only in a subsidiary fashion: they do not enter into the focus of attention. This is why we very often do not know the color of the eyes of persons whom we meet every day.

Polanyi stressed that calling the awareness of some stimuli marginal does not exclude either their being unconscious or wholly subliminal on one hand or recountable (i.e. retrievable in memory) on the other. What makes them marginal or subsidiary is the function that they play in the focal perception of the whole.

Third, we take account of some current psychological research which shows that such marginal or tacit stimuli in perception may be within my body, but that their occurrence enters into the meaning of the focally-perceived object (e.g. the phenomenon of "subception" or more generally, binocular convergence, sense of balance, etc.).

You sense that the extension of the tacit dimension is widening rapidly in these examples. Indeed we cannot inventory its contributions to awareness nor circumscribe its presence. But recognizing its role allows us to solve or dissolve many problems of epistemology. Thus Polanyi was brief with the problem of other minds: if we mean by that how we infer the existence of other persons from the observation of their bodily phenomena, the problem doesn't arise because we do not observe their bodily behavior. Rather it enters only tacitly into our perception of others.

Lastly, let us call anything which is known by relying on such tacit elements a comprehensive entity. Then we can sum up by saying that we can discriminate or identify comprehensive entities only by relying on the functioning of a tacit input, and that if we arrest that function by focussing on the subsidiary stimuli, the comprehensive entity will vanish from our perceptual field. It cannot be retrieved by the amassing of the "data-base" because not all of the tacit data may be identifiable and because the using of data tacitly requires a skill which cannot be taught.

What we have described thus far is what Polanyi called the phenomenal aspect of tacit knowing. Our concern here is with its ontological reference. A comprehensive entity, we said, is one which is discriminated by relying on the integration of tacit stimuli with focal awareness. In some cases at least, the cognitive structure reflects the structure of the entity known--e.g. when someone reads the expression on my face. I am tacitly integrating the parts of my face in expressing pleasure in seeing you, and you are tacitly integrating those parts in perceiving my expression.

Polanyi generalizes this symmetry: it is plausible, he says, to assume in every instance of tacit knowing "the *correpondence* between the structure of comprehension and the structure of the comprehensive entity".[1] The ontological structure of the entity corresponding to the dual character of tacit knowing would be as follows:

> (1) that the principles controlling a comprehensive entity would be found to rely for their operation on laws governing the *particulars* of the entity in themselves; and
>
> (2) that at the same time the laws governing the particulars in themselves would never account for the organizing principles of a higher entity which they form.[2]

Living things, incorporating parts governed by the laws of physics and chemistry, but exhibiting organizing principles irreducible to those laws, would be one example of this ontological structure; but consider another: a game of checkers.

There are rules which define legal moves and the goal of the game, and any particular game will be governed by those rules. But legal moves are not all that we perceive in watching a game of checkers. We also perceive (though a beginner may not) a personal style which circumscribes the legally possible moves and forms them into a strategy: aggressive or timid, subtle or crude, intelligent or stupid. Remarkably enough, A.L. Samuel's computer program which played champion-level checkers was described spontaneously by its opponents as not exhibiting any such qualities of play: there was no sense of personal style, no overall strategy, but only the most reasonable move in each individual situation.[3]

The conclusion that I want to draw for this section is that of the *reality* of such comprehensive entities, a reality which is unlike that of the photon because it exhibits structures which are not enfolded in the lower levels of organization. Professor Bohm is consistent in concluding that there is, with respect to the holomovement, no sharp distinction between living and non-living matter. There is ultimately only one universal order for mind and matter. I have indicated why it seems possible to acknowledge the symmetry of implicate order in knowing and known without embracing the proposal that they form only one order, but only that they correspond.

III

The issue raised here may be generalized into the question of hierarchies of order. What is the criterion (or criteria) for discriminating degrees of being or reality as we pass from photons, to atomic elements, to substance, to living and to thinking entities? Should we say that none of these is more real, has more fullness of being than another? And if not, should we mark them off in terms of being more enduring, or more independent in action, or more complex in structure, or what?

Polanyi has suggested two criteria by which to stratify or rank order degrees of reality. The first is the one already mentioned: where a comprehensive entity has parts governed by laws (e.g. of physics and chemistry) but in which those laws are constrained by boundary conditions deriving from organizing principles not themselves derivable from the laws of the parts. In this case, not all possible configurations of the parts will occur within the comprehensive entity, just as not all possible moves in a checker game will occur within the unity of the strategy of a given player. This structure of constraint by boundary conditions can be repeated at a higher level, giving rise to a series of distinguishable stratifications.

The second criterion comes from Polany's analysis of the process of discovery in science. It is that if a suspected (hypothesized) comprehensive entity is real, it will have the capacity to surprise us by manifesting itself in yet unanticipated ways. So de Broglie's wave-theory was unexpectedly confirmed by the discovery of electron diffraction. I would like to generalize this by saying that the *more* real an entity is, the more it will be capable of surprising us and the longer it will be able to do this (a thesis with theological implications). The contrast may be pointed by referring to the world of Laplace, which surely conceals no surprises, and to the study of man as Watson conceived it, namely explaining behavior in "physicochemical terms". The flat world of monism has no secrets, only problems to be solved.

The first inference to be drawn from this is that where there is stratified order their is already a universe with some kind of meaning. There is some kind of *sense* when some things are recognized to point beyond themselves, to be subsumed into a higher function, to be organized by a significance which they do not account for, even though that significance be existentially inseparable from the subsumed parts. In Aristotle's universe, the hierarchical strata form a whole which makes sense in this way even though everything that is is included in the whole. The ultimate reference point of intelligibility, the Primer Mover, although not included in the space of the natural world, is yet part of the whole, the still turning-point of the visible cosmos.

So the first kind of meaning of the whole is that of the order of internal hierarchy, articulated in a way which makes sense of the lower in terms of the higher, and whose unity is a unity of order.

The question of a second kind of meaning arises if we ask whether that order has a purposive design, whether that very order has come about in order to serve some yet higher purpose. It seems clear to me that Aristotle's answer to this question would be no, if he ever asked it. There surely *is* teleology in his world, but that teleology does not point beyond the whole.

The meaning in question here is a religious meaning. Traditionally it has been responded to by the argument from (or to) design, an argument which seeks to discern a purpose in the de facto order. In its classical form, this argument drew heavily upon Aristotelian physics and biological adaptation, and was consequently proportionally weakened by the alternative accounts offered by Galileo and by the theory of natural selection. It has recently been refurbished in terms of the so-called anthropic principle, which claims that the possible physical configurations of basic elements (e.g. internal energy levels of the carbon nucleus) were de facto restricted in the course of cosmological evolution to those which eventually made life possible.

I want to argue, however, that this form of response to the question of religious meaning (i.e. the argument from design) is transposed from and secondary to one which arises originally in another dimension, that of history as capable of narrative unity. The issue turns on the possibility of reasonably speaking of God as an agent, as an actor in history. For a religiousness untouched by the worldview of natural science, this conception presented little difficulty, but it became an acute problem for many in the modern period, and notably for the theologian Rudolph Bultmann.

Bultmann saw the Bible as pervaded by a mythical worldview in which events in this world are described in other-worldly terms. The Biblical accounts had therefore to be "demythologized", interrogated for their reality content. The task of philosophical and theological reflection is to bring this content to light by translating them into a language compatible with modern science. We cannot live in a world of natural laws, of cause and effect, and at the same time believe that Christ rose through the clouds or that miraculous violations of those laws occurred at will.

His solution was basically Kantian: I can regard e.g. a thought or decision in me as divinely inspired, even though I know it to be and perceive it, introspectively, as connected to psychological causes. Noumenally, we can say, it is a divine action, but phenomenally it is mundane. We must give up any description of God's intervention as a "breaching" or "piercing" or "disruption" of the causal chain.

There is not time to detail Bultmann's program at the length he deserves, for he stands as a man of great integrity and intelligence who sought to bring his faith and scientific convictions into concord. But apart from my reservations about that concord, there was an inescapable lacuna in his effort. Because he was a Christian, he was compelled to view the Scriptures as authoritative in a way in which, say, Plato's writings were not. The Bible, he said, has "authorized words": what it proclaims is "the call of God". So it seems that at least in this one instance, the transcendent *appears* in the mundane. Is this not a miracle, that human words can be recognized as God's action, as God's speaking?

I want to conclude by proposing a way in which we can make sense of Bultmann's problem and of his lacunna. Recall Polanyi's characterization of the way in which a higher level of meaning imposes boundary conditions on the laws governing the parts of the higher whole. Any description of that whole in terms of any of the lower levels does not disclose to us the presence of the higher level. So long as we are oblivious to the higher level, its presence will be betrayed only by what we perceive as chance coincidences: think e.g. of the way in which Freud finds the evidence of the unconscious in slips of the tongue and inclinations to step on the cracks in the sidewalk.

St. Augustine's *Confessions* is a paradigm-case of the discovery of such a unity in the otherwise coincidental events of his life, such as the post of public rhetor falling vacant in Milan. This is why God is a hidden God who is yet manifest as unseen. How can we learn to see this ultimate level of meaning of the world? Well, basically in the same way we come to see other such higher levels: by the testimony, the witness of those who see and who lead us to try to see for ourselves. If we seek perseveringly, we will find the light to see by, and the story of my life will merge into the most comprehensive of all possible stories.

University of Notre Dame
Notre Dame, Indiana

NOTES

[1] Michael Polanyi, The Tacit Dimension (London: Routledge and Kegan Paul, 1967) pp. 33-34.

[2] *Ibid.* p. 34.

[3] Arthur L. Samuel, "Some Studies in Machine Learning Using the Game of Checkers," in Human and Artificial Intelligence, ed. F. J. Crosson (New York: Appleton, Century, Crofts, 1970).

COSMIC AND HUMAN EVOLUTION IN
THEOLOGICAL PERSPECTIVE

John H. Wright

Thirty years ago, when I was working on my doctoral dissertation *The Order of the Universe in the Theology of St. Thomas Aquinas*,[1] I was struck by the absence of cosmological detail in his treatment. No doubt he was imagining a geocentric world, with spheres revolving around the earth to explain the motion of the stars and planets. But he displayed a curious indifference to these matters; he wrote for example, "The hypotheses of later astronomers are not necessarily true: for although the phenomena are saved by these hypotheses, still one need not maintain that the hypotheses themselves are true, since stellar phenomena are perhaps saved in some other way that no one has yet understood."[2] Instead, Thomas concentrated on the metaphysical principles that he judged must underlie any ordered totality. For this reason, the heart of what Thomas says about the universe is quite detachable from its medieval construct and can, I think, serve to illumine the universe as we understand it today, an unthinkably vast complex of material beings in a process of evolution that began about 20 billion years ago.

The central insight of Thomas is that the order of many things to one another is on account of their common order to an end.[3] Finality produces order, and order advances purpose and provides intelligibility. Without a common end, there is no order among acting beings. Without order there is no advancing to a goal nor is there anything that the mind can grasp or understand. This insight underlies what I will have to say about cosmic and human evolution.

I regard much of what I propose here as terribly tentative, perhaps even a bit outrageous. I worked out an initial development of these ideas several years ago when I was teaching a course on Theology of the Human Person. I talked over the scientific details with a theoretical physicist, Fr. Andrew Dufner, S.J., to be sure that I was not entirely mistaken on this score. I have also received further help on this from Dr. Robert Russell, who is Director of the Center for Theology and the Natural Sciences at the Graduate Theological Union in Berkeley. My indebtedness to the work of Prof.

David Bohm will be evident in many places.[4] It should be noted that my science is almost entirely descriptive, since my hold on the mathematics of it all is extremely meager.

I wish to advance through this topic of cosmic and human evolution in theological perspective by trying to answer three questions:

1. What is the relationship of *homo sapiens* to the universe of space and time, most particularly to the evolutionary process that has brought the human species into existence?

2. How is the whole process and the human being related to God as the beginning and the end, the ground and goal of this reality?

3. What is the distinctive aspect of the human being as a free, conscious subject, and what does this tell us about human destiny?

1. *Relationship of the Human to Space and Time*

If we view the human as the fruit of an unimaginably long evolutionary process, we must first consider the nature of that process, starting as far back as we can. It seems to me that there are several ideas, or more exactly, several aspects of reality whose interrelationship we need to explore within that process: energy, field, unstable particles, and stable particles or elemental "things." Let us take them in the reverse order.

By stable particles or "things" I understand the enduring units of mass and energy that we recognize most fundamentally in electrons and protons. They themselves appear to be the fruit of an evolutionary process in some sense. At least two reasons point to this: 1) under certain conditions in laboratories they are observed to break down into unstable particles, and, more significantly, 2) as we trace back the earlier conditions of an expanding universe we come to the very early stage where the concentration of matter and energy would not permit the enduring existence of these particles.[5]

Thus, what we call "unstable particles," i.e., energy quanta that lack mass or permanence or both (like photons and neutrinos), seem to be prior to things or stable particles, and to enter into the process of their constitution. These unstable particles can themselves be regarded as the condensations within an expanding field of energy, which by expanding permits the pure energy that constitutes it to condense into these quanta.

The notion of field is one that is familiar to us from magnetism and gravity. We recognize it first as the seemingly empty space that is in reality charged with forces that communicate energy. The initial field that we are considering here is constituted by the primordial concentration of energy that is the universe in its first instant. I wish to make an outrageous suggestion with regard to this primordial concentration: it was without any extension whatever. It was what we call a mathematical point. But because there was no space or time outside it, it could not properly be said to be anywhere or at any time. (This will perhaps not seem so outrageous, when we reflect that the universe as we know it now in its vast expanse, if it is a finite universe, is also without extension with respect to anything outside it--since there is nothing outside it. Thus, we can speak of the universe as being large or small only by comparision with what is within it; there is no absolute largeness or smallness. And if indeed it begins at a particular instant in the past, there is no prior time within which it can be temporally located. At any rate, this non-extendedness in space and time in the first instant of its being makes it possible for us, it seems to me, to conceive a relationship of dependence on a being that is not temporal and spatial in its mode of being--something we will discuss explicitly in the second question.)

The immediately subsequent moment in the existence of the universe was the violent explosion of this primordial concentration of energy, radiating outward at the speed of light, and thereby giving rise to space and time. This is the primordial energized field. In its primitive density there are no particles, only the single undivided unit comprising all the energy of the universe coming to be and enfolding the totality of its order. Space comes to be as the energy radiates outward. Time enters into the structure of space, as both measuring and constituting the interval or distance between the initial point of the energy explosion and the limit to which the energy has thus far reached in producing space. (Even now when we designate great distances through a time measurement, as for example we speak of a million light years--the distance light travels in a million years, we need to recognize that the time not only measures the distance, but in fact in some sense constitutes the distance, inasmuch as the light or energy is somehow constituting the space that it traverses during the time period that it is radiating. But this matter of the construction of space/time is somewhat peripheral to our concern here).

As the energized field expands, the over-all energy density becomes less, and quanta particles of energy appear as the condensation of this energy. But they come and go very rapidly, as the energy condenses and then returns to the field itself. These are the unstable particles.

A further stage in the evolution of the universe is reached when these fleeting, evanescent particles begin to combine and to form more enduring particles. However, these protons and electrons are themselves, at this energy level and in these circumstances, only somewhat more enduring. They readily break down into the more primitive evanescent particles and return to the energized field. This goes on over some period of time as the universe continues to expand. What is characteristic of these more enduring particles when they do appear is that they have some kind of center, in which the reality of the particle rests so to speak, and to which its totality is referred. This center enables it to persist as an identifiable unit or whole through time and to react with other such particles. As the space expands and these particles are less and less bombarded by what surrounds them, they exist for a longer time. With these particles we first encounter mass.

It should be noted that not only does the energized field produce unstable particles and these, the stable particles; but the field itself is modified by the presence of these particles, even the stable ones, for they too radiate energy and exert force.

These swarms of enduring particles, of protons and electrons, grow as the energy field expands and as more energy is converted into electrons and protons and remains in this stable form. Then, the simplest atoms begin to form, as one electron is joined to one proton to yield hydrogen. The process of constructing the elementary atoms goes on, and great clouds of gas form, highly energized, rotating, and gradually forming into galaxies and stars, around some of which planets revolve and around some of these, other satellites.

Up to this point everything has for the most part been an illustration of entropy: the diffusion of energy as space expands, loss of energy in particular locations, a continual lowering of the over-all density of energy, if we may speak in this way. But on some planets (at least on the planet earth) conditions are suitable for the emergence of a natural movement or development that builds on entropy yet goes counter to it in a way. Atoms not only unite to release energy, so that the level of energy of the molecule is lower than that of the atoms before combining, but some atoms unite to form molecules and capture some of the surrounding energy at the same time. These molecules represent a kind of negative entropy, as Schroedinger termed it.[6] These combinations are relatively unstable, but the tendency to form them is very strong when the conditions are right. Thus they continue to build, in endless forms and varieties, swimming in the great seas of water that have formed. The liquid environment allows for many interchanges and chance experiments.

All the while, as these structures of atoms and molecules become more complex, they likewise achieve proportionately more intense centers, going from electrons and protons, to more and more complex atoms, to molecules or combinations of atoms, to more highly organized molecules accumulating energy as they form. We are dealing here with organic compounds, what Schroedinger called aperiodic crystals,[7] and what Teilhard de Chardin has in mind when he speaks of the polymerising world.[8] These compounds do not endlessly repeat the same pattern, but in more and more complicated forms build up a vast array of combinations. We need to take careful note of this tendency of matter to complexify in this way. For if there were not this counter-entropic movement within matter, none of this would be taking place. It is indeed the fundamental evolutionary force. And as the external structure of matter complexifies, the unifying centers intensify; for they must exert a unifying force sufficient to comprehend the larger structure precisely as a total unity, and not just as juxtaposed particles that are somehow adhering to one another.

Another property of matter appears in the replicating molecule. A molecule is formed that reproduces its own shape and pattern from the matter that surrounds it. If the complexifying tendency of matter did not tend finally to produce the replicating molecule, capable of handling on its own structure as an inheritance, evolution would have stopped at this point of infinite variety of many molecules. Films of interconnected molecules develop which are osmotic in some way, allowing some atoms and molecules to pass through and preventing others. These films or membranes become the protective and nourishing walls of some replicating molecules, which thereby in some instances becomes the nuclei of cells, containing the formula of their own reproduction.

This, too, is another level of development of matter in its counter-entropic movement: the production of single cell living things. But the complexification of matter does not stop here. Cells combine to form colonies of single celled living beings. But these combinations begin to diversify within, different cells doing different functions, and thus another form of life appears: the many celled living being. Plant life evolves with the capacity draw energy directly from light, not just from other molecules. As they develop they draw carbon dioxide into themselves, absorb the carbon and release pure oxygen. They transform the earth's atmosphere so that nearly a fifth of it is oxygen. This makes possible the development of other forms of animal life with more complex nervous systems.

And so it goes on, living things transforming the environment and all the while moving toward greater complexity of structure and correspondingly greater centeredness or intensity of unifying energy. This is the process that Teilhard de Chardin describes in *The Phenomenon of Man* and forms the basis of the entire work: "In sum, all the rest of this essay will be nothing but the story of the struggle in the universe between the unified *multiple* and the unorganized *multitude:* the application throughout of the great *Law of complexity and consciousness:* a law that itself implies a psychically convergent structure and curvature of the world."[9] Finally, the human being appears with a unifying center that is truly conscious: it not only knows, it knows that it knows. It not only wills or desires, but it wills to will. The center is self-reflective in conscious and responsible freedom.

The human being has come to be as the fruit of the evolutionary process that builds on entropy yet goes counter to it, and takes place where the circumstances favor this development. But note, unless matter itself was already instinct with the tendency to complexify in this way, nothing would have happened. It is this insight which completes the Darwinian explanation of evolution as chance modifications and natural selection. We can speak of chance modifications only within the context of a force or tendency moving matter to form these ever more complex combinations. As Whitehead observed in *The Function of Reason*, the survival of the fittest does not explain the arrival of the fittest.[10] Without the complexifying tendency we would be at the level of protons and electrons, which are very well suited for survival.

In answer to our first question then we may say that human being, the human race, emerges from the movement of the universe to condense into stable particles, which develop into more and more complex forms having more and more intense withins: evolution accompanied by involution. These stable particles came from unstable particles, which came from the energized field which came from the unimaginably powerful expansion of the single, undivided unity compromising the energy of the universe and enfolding its order in an atom without extension.

2. *The Relationship to God: the Theological Perspective*

This primordial atom is the immediate effect of the creative act of God, the transcendent Creator. In this creative act God freely chooses to share the abundance of his reality, his being, his activity, his power to act, i.e., his ability to influence or communicate reality in some way to another. God chooses to be a creative matrix, to contain within himself as distinct from himself an immense quantity of energy, whose thrust is from God as Alpha to God as Omega. It is a thrust into the future, toward the realization of what is contained in the immense power that lies within it. It is a drive toward the divine goodness, the very reality of God as shareable with created beings. As we have said, the expansion of this atom creates space and time. Prior to this expansion, the atom does not exist in space and time, but rather space and time exist in it, as enfolded or implicated within it. The tendency to expand is the basic order to the end on account of which the order of things to one another comes to be and operates.

We human beings find ourselves in this world of space and time, we experience ourselves in this temporal process that begins from the primordial atom of energy and moves into the future that is God-Omega. Thus, we experience ourselves as emerging out of the past, willy-nilly. We experience ourselves as impelled or pushed forward into the future. At the same time, we experience ourselves as drawn into that future. The prospect of what lies ahead, the possibilities that are at hand exert an attraction, a kind of magnetism or lure for our whole being. We look ahead spontaneously, expressing thereby a kind of psychological attraction; quite literally we *face* the future.

The impulse from the past is mediated to us from all our past history, even from the past history of the whole universe. The evolutionary process that brought us into being and has borne us to this point in time and space continues to bear us into the future. The attraction from the future and into the future is mediated to us by the possibilities that lie before us, both psychologically as perceived possibilities, and physically or ontologically as the genuine possible participations in God's goodness, possible outcomes of the present situation whether we perceive them or not.

The ultimate root of both the impulse from the past and the attraction from the future is God. The creative choice of God-Alpha is the source of the push impelling us from the past into the future; the goodness of God Omega is the source of the attraction drawing us out of the past into the future. We experience both push and pull in the present.

The present is the moment of self-awareness and freedom. We choose in the present. As free agents we exercise our own responsibility at this moment which is the present. We are here cooperating in our evolution or development, in the very process that brought us into being and continues to bear us into the future. What does our choice concern? What does it not concern?

It does not concern whether we are to be in this process or not. We have no choice about whether we will move forward in time, out of the past and into the future. We cannot escape the push from the past and the pull from the future.

It does concern which of the present possibilities opening out on to the future are to be actualized. We perceive several possibilites within our power, the power that comes to us from God and is not somehow generated independently of him. It is our responsibility to choose freely among these possibilities. Freedom here does not mean that the choices are all equally attractive, that the alternatives all have the same strength of appeal for us; but that they are all seen as truly possible, in some way able to be chosen and carried through in some measure. Here we either align ourselves with the creative impulse and attraction of God or we in some way depart from it. We either accept the power of God's love seeking to draw us to fullness and developed completeness or we refuse it. We either cooperate actively in the evolution of our personalities that God intends or we fail. When this failure is seen and freely accepted as the refusal to love God, to subordinate ourselves to the magnificence of his greatness and goodness, we call this failure sin. Probably most of our choices are a bit of one and the other; but the fundamental option of our lives is at work here also. We are at the profoundest level either freely choosing God as the supreme value, or we are worshipping something else, whether we call it worship or not. We may be worshipping ourselves, or fame, or power, or pleasure, or money, or some other created reality that is distinct from God. It is this choice that determines more than any other single factor what kind of persons we most profoundly truly are.

3. *Human Distinctiveness and Human Destiny*

The distinctiveness of the human lies both in the degree of structural complexity of the human body, especially in the nervous system, and in the inner centeredness of the human psyche or 'within.' This latter distinctive aspect is not merely one of degree but of kind; with the emergence of human self-reflective awareness and human freedom the evolution of the world crossed a threshold of critical importance.

The critical nature of this threshold which is found in full conscious reflection is parallel to the critical first step in the evolution of the cosmos, when transient energies achieved stability in enduring mass/energy units. In this case the transient energies of the universe became concentrated in definite packets, able to respond to similar concentrations which were also being formed around it. Instead of simply expanding through space and time and appearing for momentary encounters with the field and with other evanescent quanta of energy, these new stable units had a center which drew to itself some share of the energy spreading throughout space and time, and fixed it at a definite point in relation to other points. Teilhard referred to this as an exercise of radial energy, where the word radial is an adjective formed from the word radius, the line which connects the circumference of a circle with its center. The energy so concentrated also radiates out from this center to react with the field and with other surrounding particles. It is here that we encounter both mass and energy as aspects of material reality.

When we come to consider the human within we meet with something altogether new, though in continuity with what has gone before. The human within its center not only arises out of the material complexity of the human body and its nervous system, at the same time stabilizing this structure and its accompanying tangential energies, it also has stability in itself, precisely as this centered within. For the human within is characterized by reflective self-consciousness and freedom. In self-consciousness the human within takes possession of itself. In knowing that I know so that my very knowing reflects back upon itself I achieve a kind of activity, which in spite of its relationship to the nervous system and bodily structure, does not rest or rely simply and directly on that structure, but is centered upon itself, rests upon itself, and in this way and to this extent is not dependent upon the without of the body. In freedom I grasp and dispose of my ability to act, reflexively willing to will, choosing among the possibilities of acting that I perceive as within my power.

We observe, therefore, how the process which first drew the random, transient energies of the exploding cosmos into stable mass/energy units wherein the radial gave subsistence to an enduring unity, has so far progressed that the stabilizing radial center is itself stabilized in itself. This radial center continues to exercise its function with respect to the body and its activity, especially the nervous system's activity, by directing from within the goal or orientation of these activities. At the same time it has functions or activities in knowing and loving which are not simply the exercise of bodily activity.

This interaction of the within and the without that we are desribing is extremely mysterious, but perhaps a clue to it can be found in the notion of the field which is both affected by the energy and mass that is within it and at the same time affects these. It has been observed that the way in which the human brain appears to store its information is by way of a field. It is not that single bits of knowledge are stored in single portions of the brain, a particular cell for example, but that they are somehow spread throughout a considerable portion of the brain, analogously to the way an image is spread throughout a holographic representation. It is this field which is also somehow directly modified by the choices and activities of the within in the fullest sense. I affect my nervous system by affecting the field that my nervous system both produces and is in turn influenced by. Thus, the direct influence of the within upon the activities of the without occurs at the infinitesimal level of nervous exchanges as these set up a field which both reflects their presence and guides their orientation or direction.

This stabilized center is able to respond not only to the surrounding objects in the space-time continuum, but precisely because of the activity within it that is radically independent of this continuum of extended matter, is able also to reach out to and to respond to the ground and goal of all reality, to God who is Alpha and Omega. As a personal center of knowing and loving the human can relate to the supreme personal center who is God, upon whose own knowing and loving the totality of all that is rests. God contains all the universe of space and time without himself being contained by it.

The Christian vision of the universe at this point acknowledges that God not only contains the universe as distinct from it, but that in the person of Jesus Christ has entered the universe as a part, has become a creature, a human being. His life, death, and resurrection have a cosmic significance; for this paschal mystery of Christ resumes in itself the total evolutionary thrust of matter, sustains and guides it, and bears it toward its ultimate goal. This thought of the "Cosmic Christ" was a favorite theme of Teilhard.[11]

This ability of the human within to function with a certain independence of matter and to relate to the ultimate source and goal of all that is so as to rest in this "ground of all reality" points to its capacity to survive the collapse of the without from which it has arisen. Though bodily death limits the ability of the human psyche to influence and be influenced by the world of space and time, this psyche is still able to relate to God as the center of centers and thus to endure beyond death. The knowledge and love which the created human center has is drawn by God into the reality of the divine knowing and loving. God-Omega draws the creature into a personal and enduring union as the fruit of love and as constituted by the inner self-reflection in and through the reality of the divine knowing and self-giving. This union of the human spirit with God and with others similar to itself, a union in vision, love and joy is the meaning of eternal life; it is the destiny of humanity as reflected in the characteristic and distinctive quality of being human.

We said at the start that for Thomas the metaphysical rather than the cosmological aspects of the universe concerned him. The relation of all things to the end grounds the relations of these things to one another, and their relationships to one another further the relationship to the end. Now the end, in the view of Thomas, is God himself immediately known, loved and rejoiced in by intellectual creatures. It is to the realization of this end that everything has been directed from the first moment of creation, when the primordial atom of energy burst from the mind and love of God into the evolving world of space and time. It is this end that finally makes intelligible all that transpires within the created universe. In the realization of this end God himself rejoices in the fulfillment of this creative love.

Jesuit School of Theology
Berkeley, California

NOTES

¹This was published in the *Analecta Gregoriana* as vol. 89, Rome: Gregorian University Press, 1957.

²*In II de Caelo et Mundo*, 1. 17. See also *S.T.* 1a q. 32, a.1, ad 2m.

³This thought recurs throughout Thomas's writings: e.g. *In I Sent.* d. 44, q.1, a.2; *De Ver.* q.5, a.1, ad 9m; *C.G.* I, 42, §7; *In XII Metaph.* 12; *De Pot.* q.7, a.9; *S.T.* 1a p., q.25, a.5; *Comp. Theol.* c. 123.

⁴I have been especially helped by two of the essays in *Wholeness and the Implicate Order* (London: Routledge & Kegan Paul, 1980): "Fragmentation and wholeness" (pp. 1-26) and "The enfolding-unfolding universe and consciousness" (pp. 172-213).

⁵A very fine presentation of the contemporary scientific understanding of the earliest stage of the universe is found in *The First Three Minutes: A Modern View of the Origin of the Universe*, by Steven Weinberg (N.Y.: Bantam Books, 1980).

⁶See Erwin Schroedinger, *What is Life?: The Physics of the Living Cell.* First published 1944. (Cambridge Univ. press, 1967) pp. 75-80.

⁷See *op. cit.*, p. 64-65.

⁸See *The Phenomenon of Man* (N.Y.: Harper and Row, 1965), pp. 70-71.

⁹*Op. cit.*, p. 61.

¹⁰He develops this thought with much wit and penetration in Chapter One. See Alfred North Whitehead, *The Function of Reason* 4th edition (Boston: Beacon Press, 1966), pp. 3-34.

¹¹See e.g. "The Christian Phenomenon," the Epilogue in *The Phenomenon of Man*, pp. 291-299.

THE IMPLICATE WORLD: GOD'S ONENESS WITH MANKIND AS A MEDIATED IMMEDIACY

William J. Hill, O.P.

Whereas the natural and empirical sciences can and indeed must resist any improper incursion of theological conclusions into their own spheres, it is become more and more acknowledged has that these disciplines presuppose some sort of world view *(Weltanschauung)* that ultimately implies not only a philosophical but a religious option as well. But it is equally apparent that theology cannot ignore the discoveries and postulates of the natural and social sciences and the humanities such as history. This is true, at any rate, if theology is conceived not solely as a confessional activity addressed to believers but also as a reflection upon a Word addressed to the world at large concerning the ultimate meaning of human existence, for which purpose it must strive to be adequate to all the data available from the empirical world. Obviously, it must do this in a critical way, applying its own criteria of meaning and truth, and not simply accepting all the claims of science in a unexamined fashion. Frederick Ferre has put this succinctly in stating, "theologians must not only find ways of incorporating current scientific explanations in their wider conceptual schemes but must also have a way of distancing themselves from the theories and models of science at any given time."[1]

The Implicate World of Modern Science: David Bohm

This being so, David Bohm's theory of an implicate world behind the explicate world of daily exchange, or more accurately stated, of an implicate world that lies within the explicate world which is the unfolding of that implicate world, cannot for the theologian fall upon deaf ears.[2] Indeed, theology finds here an ally in its own cause of surmounting a world so fragmented (Bohm's explicate world) that it appears devoid of ultimate intelligibility and loses all semblance of being truly a cosmos. It is commonplace to locate the origins of this fragmentation in the Cartesian shift of focus from reality to the concept as a surrogate of the real, providing consciousness with the certitude sought by critical philosophy. This fascination with our knowing of things rather than with realities themselves has to a degree resulted in the multiple and varied distinctions of reason being superimposed upon reality. This ratifies our preoccupation with the explicate order, with the consequence that we tend to overlook an order of existence, ontologically prior, which is one of wholeness and unity. Such a realm, called by Bohm the implicate world, accounts for the origin of forms which are "enfolded" in the implicate order and "unfolded" into the explicate order. The forms do not, in Platonic fashion, preexist in the implicate domain as changless eternal archtypes; they rather develop creatively out of that domain by a process in which the totality projects and injects itself into the explicate world. In short, the implicate realm is envisaged as a realm of creativity whence there occurs, not simply a replication of forms in new individual instantiations, but the origin of species themselves. Rupert Sheldrake (confirming Bohm's theory from the disciplines of biochemistry and physiology rather than from Bohm's own discipline of theoretical physics) refers to this as "formative causation" and as "morphogenesis," which he further describes as a process transcendent to time and space.[3] This causation is exercised by the totality projecting itself into its sub-wholes of the explicate order, somewhat as the sea is causative of its waves. Significantly, for the theologian at any rate, is Bohm's added observation that in such causation, which is in fact the phenomenon of creativity, the implicate order manifests a deep purpose and intentionality that does not always manifest itself in its explicate forms.[4]

The Metaphysical Realm as Implicate World

Clearly, at this point, a world view (*Weltanschauung*) enters the picture, at which juncture theological concerns arise. The limited interpretative framework of science is such that it cannot address directly the deepest dimensions of human reality. The question which urges itself on us is whether there is anything congenial to Bohm's implicate world in Christian thought, especially in the tradition of the Church Catholic. To approach this question in very broad terms first of all: both Bohm and Sheldrake view their theories as compatible with any one of four world views - materialism, idealism, immanentism, and transcendentalism. These views ground creativity in, respectively: matter, consciousness, the universe itself, and transcendent reality. Bohm indicates his preference for the latter two, and indeed would seem to ultimately opt for the fourth possibility when he speaks eventually of three realms of existence: The explicate, the implicate, and the ground source.[5]

The Judaeo-Christian commitment, of course, is congruent only with an over-all vision of reality as the creation of a God who transcends it in an absolute sense as its Creator. For those who stand in the tradition of the Christian West, the closest approximation to Bohm's implicate world is the world of finite being in its unity and wholeness as sought after and thought out in the discipline of metaphysics. By this is meant not the complex *concept* of being which comes at the end of the metaphysician's labor and is the product of it, but the domain of real being itself encountered at the origin of the metaphysical task. Here all things are one in the sense of that similitude-within-difference known as the analogy of being. The doctrine is well known, was rehabilitated during the Neo-Scholastic revival, and can be rehearsed here rather briefly. It still retains some illuminative power as undergirding a thought system in spite of Heidegger's claim to have overcome metaphysics, as well as the tendency in contemporary theology to replace analogical unity with a dialectically achieved identity. Analogical unity establishes itself neither in virtue of form or essence which differentiates the species and accounts for variety, nor in virtue of matter which individuates and accounts for plurality, but in virtue of being itself. The formal note vis-a-vis being is *esse* or existence (at least in Aquinas' account of things which critically speaking is perhaps the most defensible form of analogy), not as mere facticity or givenness but as ultimate actuality.[6] Outside of being then is only non-being which cannot be, but only can not be; there simply is no "beyond" to being. This exercise of existential actuality is grasped by us on the model of our intentional acts of knowing and loving (at least, these are the paradigm cases as David Burrell has indicated)[7] wherein the act of knowing precedes ontologically the positing of the known formally as known and the act of loving precedes the positing of the loved one precisely as the beloved. In other words, *esse* enjoys ontological priority over the essences it makes to be actual, as act retains priority over potency. Thus, being is to the essences that surge within it, thereby determining and delimiting it, as *cognoscere* is to the *cognitum*. As *esse* or existence, it is exercised act which cannot be thought of apart from that which it actualizes, namely essence--yet the two are really distinct so that *esse* constitutes a distinct order of intelligibility. Existing is then an act exercised by existents, which corresponds to knowing as an act exercised by knowers. This is to say that, in knowing, the intellect lives in its own intentional order, the being which the known has in its own order outside consciousness.

As to how the intellect becomes aware of this analogical character to being, i.e., this relative or proportional unity of all realities at the heart of a more obvious plurality, there is a spectrum of opinions. Most influential among these at the present moment perhaps is that of the Transcendental Thomists who, viewing the knowing subject as embodied spirit, endow it *a priori* with the capacity of "performing being," in virtue of a nonobjective, preconceptual pregrasp (*Vorgriff*) of being itself as the infinite horizon of its knowing.[8] The knowing of categorical objects (a realm not unlike Bohm's explicate world) is thus a thematization of this nonthematic oneness with being as such. However, the *a priori* element here seems a gratuitous assumption (one made by way of a transcendental method which is seeking the ontological pre-conditions for the phenomenon of knowing and loving). Without sacrificing the dynamism of intellect on which this theory rests, a more objective and *a posteriori* explanation offers itself as an alternative.

In this second theory, the human intellect grasps, at the very dawning of consciousness, the analogical character of being, i.e., the relative or proportional unity of all realities at the heart of a more obvious plurality, in an intuitional act that is at once conceptual and judgmental. The conceptual element is a laying hold of the intelligibilities offered by the various natures known; that action vis-a-vis the real order is spontaneously consummated by the judgmental element that lives out the quite distinct mode of intelligibility which is the existential actuality of those natures. In this manner, judgment attains to *esse* or existence as act, on which basis it surmises the relational oneness of what is existentially plural.

What is *explicitly* grasped in this judgmental intuition is being in one or another of its various finite modes, that is, as constricted to essences which limit it to being the actualization of some particular potentiality. Underlying that, however, is an *implicit* intuition of all beings as constituting a whole, a totality. This is due to a dynamism on the part of the intellect's act, one that is not subjective and *a priori*, as in Transcendental Thomism, but objective and *a posteriori*. It is constituted by what Aquinas refers to as the *"excessus ad esse,"* meaning thereby the mind's elan beyond essences to the mysterious and analogical intelligibility of existence. It is thus a nonconceptual aspect to intellection which latter is always at the same time conceptual.

The grounds for this awareness of wholeness are that act does not of itself bespeak limitation. This latter is rather a determination on the side of the essence which is rendered actual. Thus, implicit in the intuition of being is a nonconceptual and unthematic awareness of the unity of all finite reality in a common relation of dependence upon a Source which is *purely* actual, i.e. unconstricted by any essence distinct from act that would delimit it to its own potencies for existence. Thus, the Pure Act of Being (*Ipsum Esse Subsistens*) does not participate in anything beyond itself because beyond it lies nothing in which to participate. At the same time, it itself is the source and ground of all finite entities upon which it bestows a limited share of its actuality. All realities are one, then, in virtue of differing proportions to this one creative ground upon which they are dependent for their very being.

This notion of an intuition of being can appeal to the authority of Aquinas on three counts. First is his contention that the first note of intelligibility of which the mind is aware is that of being. He means by this that whatever is known is known first and foremost under the formality of being, i.e., we know that something is before knowing what sort of thing it is. The import of this is that apprehension and judgment coexist, each enjoying a priority over the other in distinct orders, in such wise as to issue in an implicit intuition of being. The judgment by intellect that "this thing exists" (which thing is already encountered in sensation and represented in the phantasm of sense) spontaneously gives rise to the primordial concept of being, to a grasp of finite being in all its analogicity as "that which is." This amounts to an implicit intuition of the act of being in all its analogates, not as a mere empirical given, but as the impact of the beingness of things bestowing itself upon us in that "super-intelligibility" which is something more than the intelligibility proper to the whole order of essences.[9]

Further support is provided in recalling with Aquinas and others that the faculty of intellection is itself spiritual. But intuition is the co-natural mode of knowing for spirit. This represents the express teaching of Aquinas on the manner of knowing proper to separate substances known as angels.[10] The human intellect is, of course, a faculty of embodied spirit proportioned to know the single material entity. But even as embodied it does not cease to be spirit, and conceivably this intuition at the origin of consciousness is a vestige of that spirituality. Thus, it might fittingly be called "abstractive intuition."

A third distinct argument derives from Thomas' noting explicitly that first principles are known "simpliciter" and "naturaliter," prior to all rational exploration.[11]

This encounter with being as at once plural and yet one in virtue of its participation in the Absolute, is something spontaneous and lived by the intellect; only later, reflexively, is it articulated into the concepts of metaphysical science. What it amounts to, then, is a pregrasp (or *Vorgriff*) but one that arises *a posteriori* rather than *a priori*, bespeaking a dynamism of intellect that is objective and fully cognitive rather than subjective and volitional.

The creative ground in virtue of which all reality is one, is eventually recognized in metaphysical reflection as the prime analogue in a causal analogy of intrinsic proportion. Such analogy excludes any common *logos* between God and creature. It excludes, in short, what Aquinas calls "analogy of many to one," allowing only for a "analogy of one to one."[12] The latter means that the concept involved embraces only created perfections (running a spectrum of modes in which the finite perfection is realized) which are not themselves projected onto God, but only serve as points of departure from which God can be designated but never represented. Nonetheless, it is the very substance of God that is designated in this use of analogy, and positively so, even if always in a merely relational and inadequate way. One relevant factor to be drawn from all this is that it can serve as a safeguard against "conceptualism." Thus, the importation of models from science such as physics (which are understood to be provisional and heuristic in kind) into theology, functions there in aid of *conceptual* clarification. The theologian, however, must eventually surmount the conceptual level (with all of its differences) by way of an openness to the act of existence; it is existence with all of its objectivity that lays fast hold upon us in our act of judgment, and ultimately the existence of God.

All of this amounts to a philosophical account of creation, a doctrine open in principle to rational discourse but seemingly a discourse that is only entered upon *de facto* by believers. "People do not think themselves into belief in God"; thus natural theology is not an enquiry that seeks to reach faith but one which "investigates it once it is there, seeing whether it is coherent and true."[13] It is in this sense that the ways *(viae)* of the rational mind to God, however legitimate and indispensable, are not "proofs" properly speaking, nor demonstrations in the strict Aristotelian sense, but objectifications of a prior lived dynamism towards the source of being.

At any rate, Christian thought has always viewed creation as a causality *ex nihilo*. It is not *ab Deo* as some sort of emanation from divinity with corresponding undertones of pantheism; nor is it *in Deo* in the sense that what comes into being is consubstantial with divinity as is the case in the trinitarian processions. There is, however, another sense in which the created world can be said to be "in God" in that the omnipresence of God means less that he is in all things and places than that they are all contained in him. To speak in this way is to understand "outside of God" as the realm of pure nothingness, the Void which lacks all semblance of reality and can be thought of only in a negating act. Rather, God, as it were, makes room within himself in a self-emptying act, in a divine *kenosis*, wherein the creature can come to be. What creation *ex nihilo* is meant to convey is that the origin of the creature is not by way of an act of self-realization on God's part whereby he arrives at the fullness of his divinity, but an occurrence in which God, in uncreated freedom and in an act of altruistic love (New Testament *agape)*, gives existence to the creature as its (the creature's) own autonomous existence.

What follows from this is that the divine does not enter into the definition of, for example, the human formally as human, but it does enter necessarily into the definition of the human as creaturely -- which is not to deny that the *humanum* is not fully known apart from an awareness of its creatureliness. In fact creatureliness is nothing more than this relation of dependence of the creature upon the Creator for both its coming into being and its continuance in being. The creative act consists in something of the actuality of God coming to be in the created effect. The creature participates in (from *partem capere* ; to take some formality in a partial and so diminished way) the pure actuality of the Transcendent. Thus, creation, taken actively, does not imply any alteration or change in the divine cause; it suffices that there be a transition from potency to act in the effect.

The Problem of Created Freedom

Preeminent among the challenges to theistic thought is the objection that to postulate the existence of God is to compromise human autonomy and freedom. Christian thought, almost without exception, however, has maintained that God acts in every act of each creature, including the free act of the creature made in the image of God and so endowed with freedom. God acts in the free acts of men and women and not, as it were, alongside of them. At the same time, as possessing its own act of being and freedom, the human creature truly moves itself and determines itself to become the sort of person it does become -- since, in Rahner's phrase, freedom radicaly is not simply the choice of one good in preference to another but ultimately the decision about self.[14] The divine causation in and through finite liberties -- something not unlike Bohm's unfolding of the implicate order into the explicate order -- is not, however, a determination or pre-determination or finite wills. On the contrary, through its self-determination, the human creature posits itself. It is this that constitutes the historicity of man -- possibly the major philosophical and theological emphasis of this century -- a historicity that is not to be confused with historicism. The apparent contradiction in these two assertions -- namely, that God acts in each and every act of our freedom, and that we freely determine ourselves -- is reconciled with the recognition that the divine and human agents are not partial causes of one free action but each is a total cause on a different level of causation. The sixteenth century dispute on grace, with the Jesuit Molina insisting on the primacy of human freedom and the Dominican Banez on the primacy of the divine motion, were thus the consequence of posing a pseudo-problem. The common mistake in the very beginning was in conceiving the divine and human causalities as in opposition to each other, each competing with the other. A truer starting point would recognize that the more extensive and intensive the divine causality, the closer to God in perfection will the effect be, i.e., the profounder with be its autonomy and freedom. This is confirmed by the trinitarian principle -- the more perfectly something proceeds the more it is one with that from which it proceeds[15] -- wherein the Son and the Spirit remain one in an identity of nature with the Father from whom they proceed eternally.

All of this then avoids the suggestion that men and women posit themselves in an act of absolute freedom. Human freedom is situated freedom -- situated, first, by the parameters of human nature which is a given, and secondly, by history, i.e., by the weight of the past and the opportunities of the present. These limit obviously the degree to which mankind can anticipate and so determine its open future. In short, human freedom is not rooted in sheer indetermination, even as it remains true that the human person possesses a capacity for determining the course it shall take in pursuing its destiny. As Emile Fackenheim has contended, the issue is not that of a total self-positing as in Fichte's notion of finite freedom, but that of an existential self-choice of a destiny already offered by a saving God; that is to say, what is at issue is the free appropriation for oneself of one's destiny and the means thereto.[16] To put this in more explicitly Christian terms: man's history unfolds in dialogue with God, but the ultimate horizon to that history has already been set by what God had done once and for all in Jesus as the Christ, expecially in the Resurrection wherein the "deadliness of death is overcome" (*Moltmann*). But when and how and under what circumstances that destiny will be achieved remains to be determined in the continuing dialogue with God. In more general terms: because man's being is anchored in a prior metaphysics of being, it issues in a genuine history as opposed to mere contingency and meaningless arbitrariness. It is a case of a necessary order of being tethering down a genuinely non-necessary order of history, setting the parameters within which history is made possible and given creative meaning.

In effect, this is to allow for a genuine panentheism wherein the divine is immanent to and active within each and every activity of the world. It is not a panentheism, however, in Hegel's sense of that term wherein the infinite is empty and meaningless apart from realizing itself in and through the finite, so that the dependence between the two is mutual. For Hegel, the possibility of the finite limiting the infinite is overcome by including the former in the latter -- with the consequence that Hegel finally locates the true being of things within God. On the contrary, the pure actuality of God means a similar pureness of causality, so that in causing creatively the divine being retains its autonomy, its transcendence of that which it calls into being. Creation, as the coming to be of some share of the divine actuality in the effect, does not necessitate any change or alteration in the divine being itself. Such alteration characterizes, not causality as such, but only finite causality wherein the cause undergoes a transition from being able to cause to actually exercising such causality, a transition alien to a cause that is already and always the fullness of act. In short, God is operative at the heart of all creaturely activity but without being acted upon by creatures in return in such wise as to gain something previously lacking to him.

This is the basis of Aquinas' generally misunderstood teaching that the creature is really related to God, whereas God's relation to the creature is a relation of reason only.[17] This is only intended to preclude any relation accruing to God accidentally as an increment (or diminution) of his being which would then have to be conceived of as lacking something of the perfection of being to begin with. It is not meant to imply that God does not *actually* create, know, love, redeem the world, and the like. The resulting relationship is not a mere extrinsic denomination on the part of human knowers. It is designated a relation of reason to convey that the fundament for it is something intrinsically intelligible within God, namely an actual exercise of causality on God's part vis-a-vis the creature. Indeed, in the instance of creation, the divine causality precedes any possession of existence by the creature, and so creaturely causality of any sort.[18] Failure to note this, however, lends substance to W. Norris Clarke's suggestion that ambiguity as to the meaning of "real" in contemporary thought warrants jettisoning the phrase "relation of reason" in this discussion.[19]

Neither is the panentheism suggested here to be confused with that advocated by Process Theology, in which system God's consequent nature, which alone in God is actual, is constituted by his prehending the data provided by the world. So, with Whitehead, it is equally true to say that the world creates God as to say that God creates the world (in supplying initial aims for actual entities).[20] Here, God's transcendence is merely relative in kind, consisting of his envisagement of an infinite number of possibilities. Indeed, God and actual entities of the world are seemly co-agents subordinate to the pure process transpiring between them called Creativity. For Whitehead and many of his followers, it is Creativity, not God, that is the ultimate category, a Creativity which is not itself conceived as in any sense actual. It should be noted, however, that some commentators on Whitehead -- notably Lewis S. Ford and Langdon Gilkey -- have attempted to overcome this impasse by locating creativity within God rather than as something impersonal that enjoys ontological priority over God.[21]

In a genuinely Christian panentheism, the very transcendence of God -- on which grounds our awareness of him is more an unknowing than a knowing, but a positive gain for us in that we know him to be unknowable -- is itself the ground of the divine immanence in everything that partakes of being. The total otherness of God explains that he is not present and operative in any one place but simultaneously in all places giving them their very powers of location; he is not present to any one order of being but to everything whatsoever that participates in being; his active presence is not an impediment to finite freedom in contestation with it but is the root, the very creative ground of created freedom and its exercise. Contrary to Sartre then the existence of God, far from being an impediment to freedom, is its indispensable condition.

Theological Wholeness: Unity with a Self-Communicating God

Bohm's theory of the implicate world, however, invites us to a more explicitly theological reflection. This occasions a shift in focus from nature to history, from being in its necessary structures to events in their historical contingency, and (as regards method), from science to narration. Nonetheless, this is not meant to imply the misleading dichotomy that history is the proper domain of the theologian while the exploration of the scientist and the philosopher are restricted to the ahistorical domain of nature and being. But it is meant to suggest that history, not nature, offers the most comprehensive manner of conceiving the totality of what is real -- and that revelation from God, in the full sense of the term, occurs within history, and even as history. Thus justice cannot be done to the level of intelligibility and meaning proper to the "event" by treating it simply as another kind of thing, as reducible to the entities of nature in their metaphysical oneness of being. It does imply that historical events as the products of human freedom do possess a meaning specifically their own, that they are realities of people as those who freely determines their own future.

The problem now, however, is that history is still continuing; it remains in its very contingency, still unfinished and open to the future. Of itself, then, history lacks totality and cannot be grasped as one and whole. The Christian believer supplies this on the basis of promises concretized in what God has already done in Jesus the Christ, above all in raising him from the dead. But even here, the total meaning of history is not something we already participate in but something we *anticipate* to come. The anticipation, however, is based upon what God has already done historically in Christ -- thus, we believe trustingly in the Kingdom of God already begun yet still to be consummated. What this can mean then is that "historical particularity does not do away with universality but manifests it".[22]

At any rate, here too, in history, the unity and coherence of all events is granted ontological priority over diversification and fragmentation. But this is not surmised by way of an implicit intuition of finite being as one in its differing relations to a purely actual source. Rather it occurs in virtue of sharing the life of a self-communicating God, something achieved by way of a faith-response to those disclosure experiences in which God is encountered -- whether explicitly or implicitly, whether knowingly or unknowingly, in terms of reflexive awareness. What explicit faith can discern here is a unity of all humankind (and, through humanity, of the rest of the universe) in the offer of salvation coming from God. It is a unity that does not precede the creative constitution of things in their derived but autonomous existence, but consummates that natural unfolding of an implicate world into an explicate world. It enjoys ontological priority in that God would not have gifted creatures with intelligence and freedom as natural endowments apart from this intention to consummate the world of nature with a world of grace mediated humanly and so historically. It is this insight which led Tertullian to write, *"anima naturaliter Christiana est."* It offers ground for the claim that Christianity is not one religion among the religions but rather the inbreak of God's Kingdom into the world, transcending all religions. It extends to all men and women in that it is a universal summons to all without exception to participate, not in being as such, but in that uncreated being which is proper to God alone. This participation in divine life can never be exercised by the creature as its own autonomous being derived from God, but only as a gratuitous sharing in the innermost divine life which never becomes connatural to the creature. Emphasis upon the universality of this offer of salvation from God draws attention to the fact that the distinction between the natural and supernatural order, however necessary and real a distinction, is nonetheless only an abstract distinction of formalities. In the concrete, no human exists in the natural order solely; everyone is in a state of at least implicitly either accepting or rejecting God's offer of salvation.

This approach of God to the human obviously involves his activity in the world and in history. It is important to note, however, that God remains God even in such activity; his activity remains a transcendent one and so is not to be conceived as "intervention" that is in conflict with the normal course of events. Neither is it is the case that we are simply affected by the divine power acting through these realities of nature and history. Rather, in them, God communicates to us his very self. Various theologians have referred to this as a "mediated immediacy."[23] By this is meant that God relates to us only mediately; his grace (or at least its offer) comes to us, we might say, not vertically but horizontally. But in doing so, he gives us nothing less than his own divine absolute closeness. What he mediates to us is his very immediacy. On our part, we can never encounter the divine directly but only in its mediation by creatures. But in that mediation, God himself can really come near to us. God makes himself immediately present, but only in various forms of mediacy. The immediacy at issue here is one that not only does not exclude mediation but actually constitutes it. The relationship of the Creator to the creature, then, does not suffer the restrictions that characterize relationships of creatures to one another. This notion of "mediated immediacy" avoids on one hand a supernaturalism in which God intervenes in the world, acting in a way that is alien to, if not contrary to, the normal course of events, and on the other hand, a naturalism that misplaces God's immanence in such wise as to divest him of his transcendence and view that immanence anthropomorphically. All of this is only to say that God's grace does not come to use solely within some isolated private sphere of interiority, but from within the whole of reality, including the social and the political, of which we form part.

Also, it is obvious that the unity in question here is a unity formed by a community of persons -- in imitation of that unity indigenous to the three divine hypostases: Father, Son and Spirit, who form in consubstantiality the one Godhead. This clues in the Christian, at any rate, to the truth that there is a dynamic interrelational, intersubjective, interpersonal plurality at the very core of the deepest form of unity, and one that is not destructive of that unity. All of this, of course, is acknowledged only in an act of faith. Still and all, it is grounded in human experience. The only alternative to this is a Barthian positivism of revelation which, viewing mankind as basically corrupt, dispenses with all human mediation of God's Word. The experience in question, moreover, is an interpreted experience since it is the experience of a subject existing in history and bringing the whole of his or her pre-understanding to the matter at hand. By this is meant not so much the undergoing of an experience and then the subsequent interpretation of it (though that indeed does happen too) but rather that the interpretation is ingredient in the very experience itself. The pre-understanding peculiar to the Christian, which the latter brings to present experiences, is living Tradition; here, the norm above all others, the norm not itself normed *(norma non normata)* is that articulation of God's Word that is the New Testament. Thus, there takes place what Gadamer calls a "fusing of the horizons"[24] in which the horizon of understanding out of which the text originally emerged fuses with the horizon of understanding that arises from present experiences in the existing world which also in their own way are revelatory. This introduces the hermeneutical element into theology whereby its task becomes that of reinterpreting living tradition in the light of present revelatory experiences. Thus is the *ipsa traditio* more important than the *tradita* which mediate it. Accordingly, the theologian is one who, through what the text does say, dialogues with the subject matter in order to hear, in light of present questions and concerns, what the text does not say explicitly. From a world of presently lived encounter a bridge is thrown back to the world behind the text in order to open up the way to a world in front of the text, so to speak, a world of new possibilities for mankind with God in an open future. Each generation of believers, then, must appropriate God's revelation for itself. The Christological concern, for example, is always "Who do *you* say that I am?" (Mark 8:29).

This experiencing, which lies at the heart of our enrichment through God's self-revelatory act, and which comes to us as grace not only in a vertical way (from heaven as it were) but horizontally too (since it is always mediated by creatures), is not (as previously noted) an intuition of the oneness and wholeness of things in the plenitude of being. Indeed, the question can be raised as to whether justice can be done to what occurs by describing it in strictly causal terms, wherein God is conceived as agent and mankind as patient. Even if it be granted that the 'laws' of causality operative within the universe do also hold between the universe as a whole and the transcendent (which is not a certitude beyond all questioning), and allowing therefore that divine causality is involved in the production of created grace within the souls of the justified, there remains the mysterious union with God himself brought about by way of such grace. This is a union with the Holy Spirit bestowed upon us as uncreated grace.[25] Here, beyond the explanatory categories of causality, lies the experience of a presence, i.e., of God presencing himself in the order of the intersubjective and the interpersonal -- a process suggestive of Heidegger's *aletheia*, i.e., of Being (*Sein*) unconcealing itself through the beings (*das Seienden*) to consciousness structured receptively as *Dasein*. The parallelism is merely a suggestive one since Heidegger's *Sein* is finite and grounded in nothingness (*das Nichts*).

The point is, however, that presence is always presence to another, thus bespeaking a basic relationality between conscious subjects, but not necessarily a causal relation involving dependence in being. A prime example of this (from theology) is the inner trinitarian relations which are real and intrinsic to divinity but not causal in kind. Human agents act causally on one another -- for example, in addressing another in speech -- but within that causal relationship occurs another phenomenon that is the self-communication of one person to another. This strictly personal self-donation transpires in the intentional sphere of knowing and loving.

The significance of all this is that on this intersubjective level, on the level of personhood as opposed to that of nature, it would appear quite legitimate to speak of relations between God and men as reciprocal, to allow that God alters in his freely chosen relationships to a world of free creatures. When the creature in its self-positing freedom changes, as indeed it must for either better or worse, God knows this and so his knowledge changes as regards its content, with the consequence that his love alters too in a tactical way in response to creatures. Since this is not a change in God's subjective acts of knowing and loving, acts identical with God's nature, he undergoes no increment or diminution in that nature, i.e., in the natural or entitative order as distinguished from the intentional. But there is a sense in which God is different, in virtue of having created a world of free beings, than he would be had he chosen not to create at all; he is different for having created this kind of world rather than a radically different kind of world. In short, he is different not absolutely but relationally.

To express this somewhat differently: the existence and freedom of finite beings does not compromise the infinity of God because he is already, in his eminent mode of being, everything that the creature is or can become. He is not acted upon by creatures in the sense of being passively determined by them for he is already the fullness of all determinations in the mode of simplicity. Thus, God is not in the process of becoming something that he was not previously; he does not undergo a transition from potentially having a perfection to actually acquiring it, simply because he is already omniperfect, and "outside" himself are only finite beings whose very beingness is only a deficient sharing in that fullness which he is.

Still, finite creatures endowed with freedom, and so self-determining, do introduce genuine novelty into the world. Vis-a-vis such determinations there do occur alterations in the content of divine knowing and loving. God gathers these into his own experience (if we may so speak) without incorporating them into his nature as new perfections (or imperfections). If this be so, then there is no obstacle to speaking of God as rejoicing in the values achieved by his creatures, and lamenting the disvalues -- as long as "rejoicing" and "lamenting" be understood as metaphors. This line of thinking about divinity has been argued for in detail, and cogently so, by W. Norris Clarke who sums up his understanding in these succinct but lucid words:

> God's inner being is *genuinely affected*, not in an ascending or descending way, but in a truly real, personal, conscious, relational way by His relations with us . . . . (but without) . . . . moving to a *qualitatively* higher level of inner perfection than God had before.[26]

Father Clarke's focal distinction is that between the intentional aspect of the divine consciousness and the realm of intrinsic real perfection in God himself. This present essay (more explicitly theological in character) relies rather on the contrast between nature in God conceived as absolute and the trinity of Persons conceived as intrinsic relations. These subsisting relations within the Godhead constitute really distinct hypostases or persons, but are not conceived as divine perfections, i.e., they are not understood as pertaining to the order of essence at all but to the quite distinct order of personhood. The term "person" is here taken not only in its structural or foundational meaning as that which is distinct in a relational way, (with the implication that the personal always implies the interpersonal) but also in its conjunctural or specific meaning as a distinct focus or center of consciousness, which in the divine instance means three distinct centers of a numerically single consciousness. The real distinctness at issue here rests solely upon an exercise of relationality (which Aquinas calls notional act as opposed to essential act[27]), which relationality is an eternal dynamism of becoming within the Godhead. But such divine self-communicating is not causal in kind; trinitarian language, in short, transcends the categories of causality. It is possible that the mystery can be further explored in terms of that presencing of Being which Heidegger calls *aletheia* or *Anwesen* (as earlier noted), and in which he sees the overcoming of metaphysics. What is suggested here is rather the indispensability of the metaphysical order, but as a grounding and making possible an order of genuine historicity, wherein persons posit themselves in the exercise of radical freedom (i.e., a freedom that is not merely free choice, *liberum arbitrium*, but an ultimate deciding about the self). If this be plausible, then, just as we attribute nature to God on analogy with finite nature, and existence on analogy, with the finite act of "to be," (granting the real identity of the two within God), then why can we not predicate historicity of God on analogy with human historicity? If so, then what would be understood by this latter are the trinitarian processions as an infinite, eternal becoming that cannot not be, and at the same time is what it is in uncreated freedom. Perhaps, in the end, this is only to say that the divine reality exceeds our limited conceptions of necessity and freedom. If the suggestion be a bold one, the motive in offering it lies in the fact that it enables us to allow for a dialectical relationship between the divine Persons and finite persons who truly determine themselves and so forge their destiny -- yet not apart from God.

All of this need not call God's eternity into question; it need not imply succession within deity but is reconcilable with the simultaneity of all time -- past, present, and future -- in God. Predetermination by God is something entirely different, and does compromise the genuiness of created freedom, reducing it to the mere lack of coercion or to mere spontaneity. The creature's being is its own even to the extent that it is the co-creator of its own future. God's awareness of this is not strictly a *fore*-knowledge, as if it lay in some mode of duration antecedent to time, or in an infinite, primal temporality embracing all time as one, but in eternity as a standpoint outside of time.

Conclusion

By way of conclusion, it remains to be asked what all this has to do with the theories of Bohm concerning the unity and wholeness of an implicate world. At least this: the Christian philosopher or theologians has no way of integrating such a theory into a worldview that grounds creativity ultimately in matter itself, or in consciousness, or in the universe as a whole. But no such incompatibility appears in a fourth world view which locates creativity in a transcendent source or *arche*. How God mediates being to his cosmos in his continuing creative act remains an open question still to be pursued. The history of initial resistance to theories of evolution on the part of theologians should caution us against closing off possibilities too precipitously. This may well be by way of an intermediate world (Bohm's implicate world) which if not open to empirical investigation can at least be thought of in terms of a theory and model such as Bohm's. Apart from not excluding a Transcendent, any natural, implicate world, whether giving evidence of teleology or not, should not be construed as an obstacle to mankind's shaping of his own human historical world, in freedom, yet in dialogue with the living God. Beyond this, the theory of an implicate world represents a challenge to theology to enter into a self-critical conversation with a contemporary horizon of understanding not its own.

For another, it would seem to offer an antidote to the radical deconstruction of metaphysics -- in favor of an emphasis upon the givenness of reality as recalcitrant to the manipulation by human knowing of the real.

Again, there may well be merit in its affording to Christianity the challenge of moving beyond a *participation model* (where one appropriates for oneself truth already in possession) to an *anticipation model* (where one looks forward to truth that lies ahead of us), on the basis of what is yet to unfold out of the implicate world.

Most of all, perhaps (in an age wherein belief in God is considered irrelevant for both thought and life) the theory of an implicate world can justify -- or at least contribute to -- the postulate that God as an antecedently existing ground of discourse can indeed be the foundation of theology, and that theological language about God can indeed be based upon access to the real referent of such language.[28]

The Catholic University of America Washington, D.C.

NOTES

[1] Frederick Ferre, "Science, Religion, and Experience", *xperience, Reason and God*, ed. Eugene T. Long (Washington, D.C.: The Catholic University of American, 1980), p. 107.

[2] David Bohm, *Wholeness and the Implicate Order* (London: Routledge and Kegan Paul, 1980).

[3] Rupert Sheldrake, *A New Science of Life: The Hypothesis of Formative Causation* (Blond and Briggs, 1981; J. P. Tarcher, 1982).

[4] David Bohm, cited by Renee Weber in an interview in *Revision* 5/2 (Fall, 1982) 39.

[5] Bohm, *loc. cit.*, p. 37.

[6] The Thomistic texts are well known; cf. especially, Thomas Aquinas, *Summa Theologiae* I, q.4, a.1, ad 3 and q.7, a.1.

[7] David B. Burrell, *Aquinas: God and Action* (Notre Dame: University of Notre Dame Press, 1979).

[8] Most notably, Karl Rahner, *Spirit in the World* (New York: Herder and Herder, 1968), esp. pp. 215, 224-225.

[9] Cf. the interpretation of Aquinas by Jacques Maritain, *Existence and the Existent* (Garden City, New York: Doubleday, 1957, pp. 34-35.

[10] *S Theol.*, I, q.58, aa.3 & 4.

[11] Ibid., q.79, aa.8 & 12.

[12] *S Theol.*, I, q.13, a 5.

[13] Martin Henry, "Transcendence," *Irish Theol. Quarterly* 42/1 (1976.1): 56.

[14] Karl Rahner, *The Foundations of Christian Faith* (New York: Seabury, 1968), pp. 35-39, esp. p. 38.

[15] Cf. Aquinas, *S Theol.* I, q.27, a.1, ad 2.

[16] Emile Fackenheim, *Metaphysics and Historicity*, Aquinas Lecture, 1961 (Milwaukee: Marquette University Press, 1961).

[17] Aquinas, *S. Theol. I*, q.13, a.7, corp.

[18] For a further development of this point, cf. John H. Wright, S. J., "Divine Knowledge and Human Freedom: The God Who Dialogues," *Theol. Studies* 38/3 (Sept., 1977): 450-477.

[19] W. Norris Clarke, *The Philosophical Approach to God* (Winston-Salem, North Carolina: Wake Forest University, 1979), p. 91.

[20] Alfred Whitehead, *Process and Reality* (New York; Macmillan, 1929). Free Press Edition (Macmillan, 1969), Part V, Chap. 2, "God and the World."

[21] Cf. Lewis S. Ford, "The Non-Temporality of Whitehead's God", *International Philosophical Quarterly*, 13/3 (Sept., 1973): 347-376, and Langdon Gilkey, *Reaping the Whirlwind: A Christian Interpretation of History*, (New York; Seabury Press, 1976), esp. p. 307.

[22] Edward Schillebeeckx, *Interim Report on the Books Jesus and Christ*, (New York: Crossroad, 1981), p. 61.

[23] One who makes rich use of the phrase is Edward Schillebeeckx, in *Christ : The Experience of Jesus as the Lord* (New York: Seabury, 1980), pp. 808-817.

[24] Hans-Georg Gadamer, *Truth and Method* (New York: Seabury, 1975).

[25] Cf. e.g. Paul's *Letter to the Romans*, 5:5, "the love of God is poured out within us by the Holy Spirit who is given to us."

[26] W. Norris Clarke, *op. cit.*, p. 104.

[27] Aquinas, *S. Theol. I*, q.41, a.1, ad 1.

[28] For opposing views, see Richard Rorty, *Philosophy and the Mirror of Nature*, (Princeton: Princeton University Press, 1979), and Gordon Kaufman, *God the Problem* (Cambridge, Massachusetts: Harvard University Press, 1972).

TEMPORAL INTEGRITY, ETERNITY AND IMPLICATE ORDER[1]

Kenneth L. Schmitz

Time is an evanescent reality that gives rise to a vague and puzzling concept of itself. Like many obscure and elusive aspects of our experience, however, it puts itself forward as a thing most obvious. So that, --as St. Augustine observed long ago and which has been quoted so often since,--we know perfectly well what time is until we are asked to define it. Its very familiarity prevents the estrangement from objects that our understanding seems to require. And so, in our efforts to track down the nature of time, we slide off too easily into another of our familiars,--into space. It is as though a fish were to try to explain the water it had always taken for granted and only succeeded in describing instead the filtered fluid light familiar to its fishy eyes. Perhaps, the fish would succeed a little better if it were to take for its guide its brief leaps into the evening air; and similarly, it is quite possible that my own efforts to understand time will be at their best not unlike those of a fish temporarily out of water.

The ordinary representation of time takes it to be a flow whose moments succeed one another: one privileged moment is dominant, has been preceded by others no longer present, and is to be succeeded by others yet to come. Taken as an initial and familiar representation, such a characterization of time is not incorrect. It recognizes the most obvious feature of time and represents it as a distinctive sort of succession: past moments were "nows" but are no longer, future moments will be "nows" but are not yet, only the present is "now." With such a representation, however, thought about time has scarcely begun, and several distortions must be avoided in proceeding to think further about it.

The possibility of distortion arises with the reduction of time to space: *the spatialization of time*. To be sure, it is possible to represent time as a sort of indefinite line, each point of which stands for a different moment. The spatialization of time is not in itself a corruption, and has proven endlessly helpful in theoretical constructions. It is necessary to remember, however, that such a translation of time into space employs an analogy and cannot be taken as a literal representation of time, since every analogy also contains a dis-analogy. If time may be likened to space in some respects, it cannot be likened to it in others; and taken in its integrity it is not equivalent to space at all. Indeed, if the representation were taken literally, the *temporal* nature of time would be lost; for each point on a line is solidly there along with the other subsistent points, whereas in time itself each moment reigns for its brief day only to pass away in favour of its successor. Time has an ingredient of negation in it that is not taken into account by the representation of a line of equally subsistent points. Nor can we improve upon the schema by representing time as a moving point along an imaginary line, for that would reduce time to a single actual point, whereas even the "now" is, as we shall see, an internally differentiated structure. Moreover, the deficiencies of this initial representation would be further exaggerated, if we were to proceed from the spatial representation of time to its *atomization*. Under the influence of an extreme and primitive mechanism, and on the nominalist supposition that unity is ultimately reducible to absolutely simple, utterly incomplex units, the points would then be imagined to be entirely discrete from one another and self-contained. Such an atomizing of time into unrelated "nows" would make of each present moment a miracle without precedent and without consequence,--linked, if at all, by our mental connectives alone, and therefore having no accountability in reality. Time would then be simply a construction we put upon reality.

In the *Physics*, Aristotle started with the ordinary understanding of time *(chronos)* and arrived at his famous technical formulation of its nature: "For this is what time is: the number of precessions and successions in process."[2] This has usually been rendered: "Time is the measure of movement in regard to before and after;" i.e., it is enumeration in respect to what occurs successively. In the formula, *arithmos kineseos*, special attention is given to the ordinary understanding of the sequential character of time. The number of which Aristotle speaks, however, is the measure of movement and presupposes its continuity.[3] For that reason, the formula must be understood in its unity and its integrity; because time is neither simply *kinesis* (movement) nor simply *arithmos* (enumeration), but rather *arithmos kineseos*.

Time is not movement. Not only do we tend to "spatialize" time; we also tend to equate it with movement. To begin with, although time is closely associated with ordinary movement, it is not simply identical with it. I intend to show that our intuitive sense of the irreversibility of time is well-founded, whereas ordinary movement can be reversible,--waves advance and recede, tides rise and fall, seasons wax and wane. More or less the same situation may return tomorrow, but *today* does not return. Geological processes may reverse themselves,--deserts come and go, mountains heave and erode,--but the geological *record* does not reverse itself. What happened millions of years ago may happen again, but the *happening* doesn't happen again. There are, of course, movements that to all appearances are irreversible, such as growth and decay; for when growth from acorn to oak "reverses" itself in the decay of the tree, the original acorn does not reappear.

Suppose someday that we could reverse the process and reconstitute the original acorn, so that it would be indistinguishable in composition, structure, function and appearance from the original. Still, in actuality it would be reconstituted; that is to say, even though it were (by hypothesis) the original acorn, it would be that acorn *qua re*-covered. The retrieval as such would not be preserved in its composition, structure, function and appearance, since--once again, by hypothesis--they would not be discernibly different in the reconstituted acorn from what they had been in the original. Nevertheless, the acorn would not be the same in one respect at least; and that non-identity would be preserved if only because the recovery would be registered by the difference in the *temporal condition* of the reconstituted acorn. I should almost want to say: "by the temporal number of the reconstituted acorn." In some sense a thing *is* its past and in some sense may be said to *"include"* its past. Now, the acorn--by hypothesis indistinguishable from the original which it "once again" would be--would have a past that the original acorn did not then have.

The subjunctive mood in which I have cast the foregoing counterfactual "hypothesis" is meant to remind the reader that I do not intend it as a proof. It is only the analysis of an intuitive certainty regarding the peculiar difference that time makes. By hypothesis, nothing physical would be changed; yet a difference would remain which would have to be accounted for. As the number of movement, time is not *separable* from it; nevertheless, it is still *distinct* from movement, for it is precisely time which makes the

101

difference that takes into account the real movement of *re*-constitution.

It is not easy to catch the sense of "re-." Let the term "reconstitution" stand for the physical result of a change. Thus, for example, in verifying someone else's experiment, suppose that we begin with a chemical substance, break it down and recombine it; we will have "reconstituted" that chemical compound and verified the analysis. But, even though the elements present in the "reconstituted" compound are the very elements of the original substance, nonetheless, there is a certain ambivalence as to whether we have or have not "recovered" the "same" substance through the "reconstitution." The difference is subtle but not unimportant. In physical or material terms, we are entitled to say that, indeed, we have recovered the same substance by means of reconstitution, since even the elements are the same. Indeed, according to the example, they are not only the same in kind, but are the very elements with which the substance was originally composed. A difference remains, however, that does not show up as a material difference in composition, but which is rather a certain "non-material increment" that is somehow characteristic of the new state of the substance, making it "different" from its original state. In this example, the "re-" is meant to identify the original state: we have "recovered" the substance, but not the original state. And so, just for that reason, the reconstituted substance is in one respect not the same and in another is: it is the same substance but it is that substance found *again*. Now, that "again" explicitly refers us back to a state prior to the change, prior to the process of break-down and (re)constitution. So that the prefix "re-" does not simply, nor even primarily, denote "identity of material composition"; rather, it bonds two states of the substance in a distinctive, determinate ordering: that of beginning and following, then and now, early and later.

In any event, if I have understood Professor Bohm correctly (and more generally, if I understand the implications of entropy), then the hypothesis of a return to what is in *all* physical respects *simply identical* with an earlier state of affairs has no real possibility (i.e., no possibility in the real world). This is because prior movements impact upon succeeding movements in such a way as to produce an enfolded result that (taken as a whole) is new. Of course, this introduces another difficulty; for if the hypothesis of reversibility has no real possibility because change is not strictly reversible, and if time is not strictly reversible either, then (even though it does not follow with logical necessity) we might be tempted to conclude hastily that time and such movement are indeed the same. Further reflection, however, will show that time is not thereby equivalent even to irreversible physical movement; for time measures such irreversible change in an irreversible way all its own. In a moment I will renew the attempt to articulate the distinctive character of time, even while I maintain its inseparability from movement; but we must first turn to another possible misunderstanding.

Time is not simply number. If time is not movement, neither is it simply numbering. First of all, numbering considered as reckoning is reversible. What has been added can be subtracted, whatever divided multiplied again.[4] Time is a *number* in the sense of a sort of numbering; but it is not simply number, it is not number without further qualification. It is just this qualification that I seek to clarify. I have already said that time is a distinctive sort of number in that it is the *measure* of movement. On the other hand, it is not just any kind of measure of movement. After all, movement admits of many different kinds of measure: of distance, velocity, force and figure; whereas time measures none of these. It measures duration.

The term *measure* is by no means obvious when it is said of time. Usually, when we say that time is a "measure," we have in mind clock-time or calendar-time, i.e., chronometry or chronology. Now, there is an obvious element of convention in the units by which we measure time; for we call a "day" 24 hours and not dawn to dawn, a "month" 28 to 31 days and not the waxing and waning of the moon. But there is more to temporal measure than our conventions. To be sure, when we reduce time to chronometry we can introduce discrete units at our convenience and determine the unit of time by convention. Nevertheless, even in chronometry and chronology we have usually chosen some natural motion as the basis, such as that of the moon, the sun, the stars, or the speed of light; or in measuring social change, we have chosen the death of the monarch, the birth of a religious figure, the recurrence of a feast, or some other unit of social significance. And so, the clock and the calendar are not simply the expressions of our conventions, but are rather themselves measured by the movement of something else. In this sense, then, the device by which we measure is itself measured by that which it measures. What, then, does it mean to "measure" when we speak of time? There is, no doubt, a certain reciprocity in such a relation of "measuring." It might be more accurate, therefore, to speak of the clock as expressing in more or less conventional units the duration that is *its* measure. Thus, we speak of a clock running fast or slow. The measure appropriate to time proper, then, is not the reckoning itself, taken simply as the calculation of units, but the basis of that counting insofar as that basis is not simply movement but the measure of movement.

So far, then, I have indicated that time is neither movement nor number alone. In speaking of time as a "measure," therefore, we ought not to identify temporal measure simply with the reckoning that expresses time, just as we must refuse to identify it simply with the movement that lies at its base. But this suggests that quantification may not be the primary nature of time at all; and, indeed, we see this confirmed by a certain qualification included in the traditional formula. For the formula says that time measures movement in terms of "before and after." And so, it is not enough to say that time is the measure of movement. The tag-phrase, "with respect to before and after," is an indispensable qualifier. It introduces a qualification into our understanding of time as a measure; for temporal measure does not determine primarily "how much," but rather specifically "how," i.e., "in what way?" Time belongs, then, primarily and fundamentally to qualitative rather than to quantitative measure. Indeed in the *Science of Logic* Hegel insists--correctly I think--that measure in general *(Mass uberhaupt)* is only secondarily quantitative and in essence always rests upon qualitatively determinate modes of being.[5] Bergson, too, insisted upon this so strongly--against the mechanistic quantification of time--that he denied any likeness of *real duration (la duree)* to number, seeing duration rather as a "qualitative multiplicity" that remains quantitatively indivisible. He therefore sought to found time in an intuition that rejected all conceptual rationality as "cinematographic."[6] That price, it seems to me, is too high to pay in defence of the qualitative character of time, and one that I think is unnecessary. I should rather say that time is not *primarily* quantitative, rather than to say that it is not really quantitative *at all*. Nonetheless, it is to the qualitative nature of the differentiations in time that we must look if we are to determine its primary, proper, and fundamental nature.

The quality of time. The most obvious temporal differences are exhibited in the succession of the moments: their "before and after"; but this succession also exhibits their inter-connectedness. Aristotle remarks that "the present is time's continuity, since it holds past and future together and is their common boundary as the beginning of the future and the end of the past."[7] He also points out that "time depends upon the present both for its continuous and for its discrete character," just as the point "at once holds the line together and divides it."[8] In Aristotle's sense of the term "principle," then, the present is a constitutive principle of time, since it provides the *differentia* between the other moments of time, even as it connects them. Like all Aristotelian principles, it produces contrary results within its own order: in the order of duration the present both connects and divides the other moments, even as in the order of space the point both ends and begins a line.[9] Undoubtedly, this emphasis upon the present as actual is in harmony with Aristotle's over-all emphasis upon the primacy of the actual in other orders as well; thus, in the order of rational process the actual is *entelecheia*, in causality it is *energeia*, and in structure it is *eidos*. The inter-connectedness of the moments of time is to be accounted for, then, in terms of their actuality (present) and inactuality (past, future).

I

While acknowledging and accepting the Aristotelian analysis of time as far as it goes, I have been impressed with the more recent contributions of Kant and Husserl to an understanding of temporality. Drawing upon them, --although pressing the analysis to purposes other than theirs and therefore introducing distinctions not drawn by them,--I should like to articulate what I take to be the inner complexity of time. It will require a recognition of three "dimensions" in the number that is time: for there are the *moments* of time, the *modes* of time, and what I shall call the *phases* of time.

The phases of time. Edmund Husserl[10] has provided an analysis of time that exhibits the interconnection of the phases of time, which through their mutual internal relations together constitute time. He showed how the phases involve each other in time as we experience it. Beginning with the lived experience of the flow of time *(Erlebnisstrom)*, he recognized that it always includes a receding past and an encroaching future. But he did not merely observe that one moment follows upon another; rather, he showed how the present is already bound up with its immediate past and its immediate future. So that the *now* *(Jetzt)*, --as a unit of real duration and not merely as an abstract point on a line, nor even a calculated amount of time--is not merely one simple moment, the present. The *now* is, rather, a sort of triad of duration; for in it that which is just-passing-away is still conjoined with the actual present, even while that which is not-quite-yet is also conjoined with the same present. Husserl used the spatial imagery of a *horizon*, but it is not to be understood as a merely static circumference; rather the imagery is that of the indefinite limit as we experience it in perception, i.e., as ever-receding. In other words, it is a dynamic limit.

The horizon of the *now* embraces or encompasses what I have called the three "phases" of time; and I will shortly distinguish between the phases and the moments of time. It is enough here to notice that time is constituted of an unbroken retention *(Vorhin)* of the past in the *now*, and of an unbroken protention *(Nachher)* of the future in the same *now*. It is important to stress the unbroken immediacy of the phases of retention and protention with the present, since each is a special mode of past and future. Unlike the recollection of events over and done with, or the anticipation of events in the distant future, retention and protention are the immediate neighbors of the present within the complex *now*. It is as though there is a certain overreach of the present back into the past (even as the past emerges into the present) and an overreach of the present into the future (even as the future intrudes upon that same present). He insists, too, that retention (the past as immediate to the present) and protention (the future as immediate to the same present) are not merely empty forms. If time is taken as the measure of something real (i.e., as experience by Husserl, or as schematised content by Kant or as natural change by Aristotle), that reality is the very content of the moments of time. Moreover, it is important to note that Husserl is not simply restating the continuity and trans-atomic relation of the moments of time to one another, as Aristotle had already done. He is showing that the moments intrude as phases upon each other, that they interpenetrate one another and thereby co-constitute one another in a single temporal unity.[11] Nor is this interpenetration simply at the level of the *meaning* of "past," "present," and "future." Rather, in their very *reality* as phases, the moments are inseparable from one another. Their mutual implication is not simply one of logical definition, but is constitutive of their very being.

It is just here that I think it is required both to distinguish a phase of time from a moment, and to also recognize their basic identity. Aristotle spoke of time as a continuum of nows coming into being and passing away in favour of successive nows. Husserl showed that the *now* is itself a complex unit of time by distinguishing the unbroken and immediate relation of the "retended" past and the "protended" future, even while he recognized their conjunction with the present in the *now*. The *now* is the principal unit of time, and I have called the immediate or "retended" past and the immediate or "protended" future the "phases" of time insofar as they are aspects of the *now*. The present is also a phase of the *now*. Even more precisely, perhaps, we might speak of the present as the positive "phase" of time, i.e., its manifestation or appearance, and the immediate past and immediate future as "anti-phases." The latter term brings out the inherent inactuality of past and future, an inactuality that consists in their disappearing in favour of the present phase in the *now*.

The moments of time. On the other hand, if we have in mind a past that has already disappeared, rather than one that is just disappearing, then we can speak of a past "moment" rather than a past "phase." That past moment is not merely the past as phase, but is the whole complex *now* as a unit of duration in which all three phases of that *now* have become inactual, have disappeared, have come into the status of pastness. Something analogous can be said of the future, distinguishing the future as phase from the future as inactual moment. Here, unlike the past in which the phase becomes the moment, however, the relation between the future as moment and the future as phase is quite the reverse, since the future moment becomes a phase of the *now*. With regard to the present, the relation between phase and moment differs yet again from the relation of moment and phase in past and in future. For the present *qua* phase of the *now* is that which *qua* moment gives the value of actuality to the whole present *now*. In a word, there are very significant differences in the relations between the phases and the moments in regard respectively to past, present and future. To articulate those differences further would lead us beyond the scope of this paper; and so I will illustrate the implications of the distinction from now on only in terms of the past.

The continuity of time. Having made the distinction between the past as phase and the past as moment, however, it is important to recognize that it is the same past that is both phase and moment. As a phase within a *now*, the past is that very "coming to be no longer" that, when it has become a moment in its own right (i.e., a past *now*), suffuses all the phases of that *now* with its "no longer." This identity of the past-as-phase with the past-as-moment is the temporal quality of the continuum that Aristotle insisted lay at the basis of temporal measurement; that is, it is the temporal account of the continuity of movement. It is that continuity expressed in terms of the values of duration.

Within their togetherness in the complex and internally differentiated *now*, the phases themselves in their becoming become the moments of the flow of time. The moments retain their qualitative distinction from one another, but not at the price of their fundamental identity with their correlative phases. This fundamental identity of each moment with its phase, and the unity of the phases within the *now* is *the ground of the very continuity of time itself*. And even more, the same fundamental identity of each moment with its phase is *the ground out of which the flow of the moments arises*. That is to say, it is through the internally related phases of time that the moments are themselves internally related and continuous. The sentence needs to be parsed: Because the phases of time are related internally by a doubly internal relation--i.e., within the integrity of the now which they constitute, *and* within the identity of each phase with its proper moment, --the *nows* (i.e., the moments) are themselves internally related and continuous. This continuity is not simply the continuum of movement; it is the peculiar continuity of duration.

The transiency of time. But time has its peculiar discontinuity as well. For within the triadic *now*, the present bears the value of what is actual. Moreover, as Husserl insisted, the "actual *now*" must retain a persistent *punctuality*. This principle of temporal differentiation is rooted in the present, because it is its presence that distinguishes the actual moment (the actual *now*) from the inactuality of past and future moments (inactual *nows*).[12] This principle of difference or discontinuity does not, however, simply separate one moment from another. For the actuality of the present moment (the actual *now*) is not such that it excludes the inactuality of the other two phases, since they are not *simply* inactual, any more than the present phase is simply (i.e., through and through) actual. This is the transiency of time, the negative factor within the present. For the present, too, is passing away insofar as it is the principal phase of the present moment (the present *now*) that is itself becoming a past *now*.

It is not easy to describe the way in which the phases and moments are inactual. I have already said that a phase is inactual in a somewhat different way from the way in which its correlative moment is inactual. For the phase past as is inactual as disappearing, whereas the past as moment is inactual as having disappeared. On the other hand, the future as phase is just appearing, whereas the future as moment has not yet appeared. Nevertheless, it is not enough to say, even of moments, let alone of phases, that they simply are not: the past no longer at all, the future not yet at all. It is not enough to say without qualification: there is no past, there is no future, there is only the present. But, on the other hand, if they are not simply inactual, neither is it true to accord to them some small amount of complete actuality, as though they are mini-beings. Past and future are inactual, each in their own way, but those ways are the ways of *duration*, and not the ways in which *things* may be inactual. There is the unbroken immediacy of the phases that fill the dynamic horizon of the *now*, and there is the succession of *nows* that makes up the flow of time; and so, the character of the actuality and inactuality of time is that of *transience*. The blend of actuality and inactuality appropriate to time is exhibited in the very coming to be of the *now* · its coming-into-being-from... and its-passing-away-into... For the very character of the *now* is that it is ever coming into being and passing away. It is this very transience, however, that ensures that time is the measure of a kind of real being, i.e., of real movement in which things come to be and pass away.[13] The horizon of the *now* is itself temporal, is itself transient, for the *"now"* is just the name given to time in its momentary integrity, i.e., in the togetherness of its phases and the continuity of its moments.

The asymmetry of time. There is, then, a certain qualitative structure that is essential to time, and it includes the immediate internal relationship of the moments to one another through their phases: the inactuality of two of the phases (past, future) grounded in the actuality of one of them (present). Thus the structure is asymmetrical in several senses. First, the past and future may be said to participate in the actuality of the present, i.e., *qua* phases, and so they are not simply inactual just because they are not yet entirely past or wholly future; they are, rather, becoming past and coming to be. Second, the inactual moments differ from one another in the way in which they are inactual. I have already remarked that the past passes from phase to moment, whereas the future passes from moment to phase. Similarly, under the "presidency" of the present in the *now*, the past passes from actuality to inactuality, whereas the future passes from inactuality to actuality. It is only if we wholly abstract the moments from the dynamic structure of which they are constitutive and separate each out for our inspection that we can speak of both the past and the future as wholly inactual. In such rarefied air they collapse into identical nothings, differing only in empty names. On the other hand, if they are to retain their difference from one another it must be by retaining their hold upon the structure from which our propensity for abstractions has torn them. In real time they play out quite distinct roles. This difference discloses a third asymmetry and touches upon the essential and unique nature of time. For in and through their phases the moments interpenetrate and display a necessary relationship that is unique to time: it is the irreversibility of their order to one another. The extreme reduction of time to space can provide us with the illusion of reversing time, but the qualitative structure of time discloses the real impossibility of re-ordering the moments of time. We can, of course, describe time retrogressively, as Merlin in E.B. White's trilogy is said to be able to predict the future but never remember the past. Still, such a bizarre description gains its mad surprise only on the firm basis of the qualitative unalterability of the dynamic thrust of time: time past, time present, time future. So that, even in transience, even in the succession that is proper to real time, a certain necessary order discloses itself.

The modes of time. But there is more to time than its moments and phases, and I have found Kant especially suggestive in regard to this "more." We need not accept Kant's restriction of knowing to the realm of appearances in order to learn from him about the further complexity of the structure of time. It is not, however, in his Transcendental Aesthetic that I find these indications, but in the Transcendental Analytic, and especially in the Analogies of Experience. His brief yet comprehensive consideration of time[14] is especially suggestive in that he does not confine it to the moments of time, --past, present and future. Of fundamental inportance are the "modes" or "features" of time. Like the moments, they too are three in number: permanence, succession and simultaneity.[15] These modes or features of time are qualitatively diverse. Permanence is not succession, nor is succession simultaneity, nor do they differ from each other in number. Indeed, for any given magnitude of time, they have the same value. Yet each fills out the whole of time in a way in which the moments and phases do not. To speak of them as "parts" of time is even less satisfactory than to call the moments "parts." We hesitate to call the moments "parts," because two of them are inactual, whereas the primary meaning of "part and whole" is drawn from spatial imagery and first of all designates a static composite.[16] But even if we were to say that the moments are "partial," they would be so in a way in which the features are not. For, as we will see, there is a sense in which the whole of time is permanent, the whole of time is successive, and the whole of time is simultaneous; whereas we cannot say that the whole of time is past or present or future, though we can say in a certain sense that it is present or past or future as a *whole*. The features of time differ from one another, nor can they be simply equivalent to time itself, but they are not parts of time; they are *integral* or holistic features of time, i.e., they are "modalities" through which time presents itself as a whole. Now, not only are the moments and phases of time bound together intrinsically in a necessary order, but the features of time are also bound together internally in the totality of the time-structure. This becomes clear, when we consider each of the features in relation to the whole of time.

First of all, as to the feature of *succession* I have already acknowledged the necessary order that prevails among the moments in their very transiency, and have suggested the grounds for both the continuity and the irreversibility of time. The continuity and irreversibility are realized through the modification of the past by which it passes from being a phase to being a moment, in contrast to the future which passes from being a moment to being a phase. Once again, special analyses would have to be made for the present and the future; but what has been said should suffice to establish the characteristics of continuity and irreversibility. And so, we can understand why temporal succession is not merely the connection of simple units, but rather the internally related sequence of intrinsically differentiated complex *Nows*.

The second feature of time is *permanence*. Permanence is ordinarily and primarily said of that in things which somehow persist and remain throughout the change. For the permanent is not the static, which is either a real state of rest *(stasis)*, or a mental representation from which the consideration of time has been excluded *(ens rationis)*. The permanent in its ordinary meaning is just that which abides, literally: that which dwells throughout (a change of state). Now, just as a substrate undergoing change both does and does not change--for example, a person in joy and then in sorrow both is and is not the same person: not the same because changed in state, but yet the same because not annihilated in substance; or again, the same tree is at one time sapling, at another forest giant, --so, too, the integral triadic time-structure is a measure that prevails throughout change. The parallel is not exact, of course, since, if it were, the permanent in time would be in all respects identical with what is subsistent (we might say "residual") in things. On the contrary, whereas matter in the sense of the substrate is passive relative to the change of states, time resembles more the active and persistent dynamism *(potentia activa)* which Aristotle discusses in the *Metaphysics*.[17]

It is important at this point of the analysis to appreciate this second kind of permanence. For, if the analysis is correct, then time does not only include permannence as *one* of its three features; in addition, the structure of *time itself* is permanent in its own way. That is to say, the qualitative temporal structure as a whole survives the transience of its moments; so that the enduring unity of the whole time-structure exhibits a distinctive kind of permanence. The ordinary sense and the original reference of the term "permanence" is to *things*, of course, or precisely, to that in things which undergoes change and survives in the result. Yet, when we open our experience up to reflection and analysis (as I have just done), it becomes manifest that the temporal structure itself "prevails" in its own way as the durational measure of movement. By "a distinctive kind of permanence," therefore, I mean that the permanence of the time-structure is not the same kind of endurance as the residual substrate which remains during the changes undergone by physical things; and the reason for the difference is that time is not a *thing* at all. It follows, then, that this new meaning and reference to the time-structure itself is not simply a second meaning of the term; rather, it is a second order of meaning, a second-order meaning of "permanence" (*permanens*^{2}). Unlike the ordinary meaning of the term, which signifies a relation between *things* and their *states*, this second-order meaning refers to a sort of meta-relation between the integrity of the *time-structure itself* and its *moments*. This second-order permanence consists in the cohesive unity of the time-structure itself insofar as it persists in and through the transient "flow" of its moments. This does not mean that, inasmuch as time persists throughout change and in a certain sense "presides" over it, the time-structure is indifferent to change; on the contrary, time is engaged with change as its durational measure. As with the permanence of things, so too, this distinctive kind of permanence does not render the time-structure static. The structure is both the same, because it does not alter its triadic constitution, its modality or way of being; and yet it is not the same, for it is never the same time, being at any given time always earlier or later than the same temporal structure at any other given time.

Now, this second-order permanence--i.e., the permanence of the triadic time-structure taken as a whole (*permanens*^{2}) --is not simply a curious mental product of analysis; it is a genuine characterization of a real modality of time itself. The point may bear repeating: Without the feature of permanence in time (*permanens*^{1}) there could be no temporal measure of the change of things, since there would be no sameness in duration upon which to base the measure of the differentiation that is ingredient in the very nature of change.[18] Without permanence in time (*permanens*^{1}), then, there could be no time, since there could be no succession of moments, but only substitution. There would be only the discontinuous appearance of instants without precedents or consequents. And so, too, while it is important to recognize that time must include the feature of permanence within it *(permanens¹)*. It is important also to recognize that the temporal structure itself does not alter,¹⁹ --for if it did, time would become something else or cease to be altogether; and this latter resilience is not the durability in things but time's own distinctive permanence *(permanens²)*. Time itself abides, then, by outstripping the momentariness of its own moments, surviving their transience; and this is *its* permanence *(permanens²)*, as distinct from the permanence of things *(permanens¹)*. In a word, such a permanence is not the ordinary permanence of things, but the permanence of time itself; is not a "part" of time, but is rather a feature, modality or aspect of time taken as a whole. It is time in its dynamic self-insistence.

We arrive, thirdly, at *simultaneity*. According to the ordinary meaning of the term, things are simultaneous when they are co-existent, i.e., when they share the same moment of time. But it is also true that the features of time--permanence, succession and simultaneity--are themselves simultaneous with each other. This is a new sense and reference, a second-order simultaneity *(simul²)*. For it is no longer the simultaneity of *things* co-existing at the same time, --in the present, past or future; it is rather the co-existence of the *features* or *modes* of time with one another. Again, it is not the co-presence of things in a particular moment of time, but the co-presence of the modes of time in and throughout *time itself*. In sum, the integrity of time consists in the permanent co-presence *((permanens et simul)²)* of the modes of time in the succession of moments.

There is, however, yet a further kind of simultaneity ingredient in time. It is the co-presence of the *phases* of time to one another, a co-presence which was shown earlier in the analysis of the complex and differentiated *now*. Insofar as the past and future are co-constitutive with the present in the very constitution *of* the *now*, so far are they co-present to one another *in* the *now*. This is yet a third sense of simultaneity *(simul³)*. for it is neither the co-existence of *things* in a moment of time *(simul¹)*, not the co-presence of the *features* in the very make-up of time *(simul²)*, but is rather the co-presence of the phases of time in the qualitative structure of time itself *(simul³)*.

114

There is undoubtedly more to be said about time, and I can only hope that I have not added to the things that have been mis-said. Time is a magnitude and therefore quantifiable; but it is not merely nor even primarily a quantity. It is neither reckoning nor movement, but the measure of movement; and not just any measure, but the measure of succession. It is not, however, simply a flow, but has, rather, its own complex qualitative structure and dynamism. It is a flow of *nows;* but each *now* has an intrinsic structure of modes and phases that are at once permanent and simultaneous. The continuity of *nows* is brought about both by the permanence and simultaneity of its modes and phases, and by the interchange between the three phases and the correlative moments of past, present and future, i.e., by the transiency that is time's most obvious trait.

Can we say, then, that the explicit and actual *now* is the tip of an implicate order of temporality in which the whole of time is enfolded in each of its instants? The manifest succession is the actual procession--not of simple atomic moments--but of triadic and differentiated *Nows;* but the order is that of the permanent co-presence of the modes and phases of time. To be sure, this is not the implicate order of movement, since time is not simply movement. We ought to expect important differences, since the foregoing analysis of time rests upon the distinction of time from both number simply taken and movement simply taken. Time is neither numbering nor movement, but the measure of movement *(arithmos kineseos)*. To resume: It is not mere numbering, since there are non-temporal enumerations.[20] Moreover, while we can set the clock back, i.e., we can re-number, we cannot set time itself back.[21] On the other hand, it is not movement alone, since at least some change is reversible; so that the generic character of change permits reversibility. Nevertheless, as the measure of change, time is not unaffected by movement even though it is not identical with it.[22] It might be said that, after all, I have been talking about change all along. This might be so; but I have been talking about it in terms of the duration that measures it, and not in terms of other aspects of change, such as alteration of state, increase and decrease, development and degeneration, emergence and destruction, and other dimensions of movement. It has seemed to me that time deserves analysis in its own terms.

The upshot of the analysis is this: time is constituted of three modes and three moments; their interrelation comprises the unique structure of time, a structure that is not comparable with any other. The mode of *succession* implies a series of nows whose very "boundaries" are temporal, and whose internal constitution is the complex intrinsic permanence and co-presence of the moments and the features of each other. The mode of *permanence* is also uniquely temporal, for it must be distinguished from that of subsistence in the order of things, from the substrate that both changes in its unchange and does not change in its changing. Finally, the *simultaneity* of the modes and moments of time must be distinguished from other forms of togetherness, since the parts of space are contiguous, and elements come together to form a compound; but the simultaneity of the modes and phases of time is a durational togetherness, a togetherness not strictly in time as we usually mean it, but a togetherness of the very components and necessary ingredients of time. They are together "in" time in the sense that they *are* time; they "make up" this unique, differentiated and complex dynamic structure of temporality. It seems to me that this unique integrity of time manifests in its own way an implicate order.

II

There remains the question whether the analysis of time can and needs to be carried further, whether a kind of implicate order is present in time which might contain the seeds for an understanding of eternity. At first glance, eternity seems to be one of those notions that we can safely set aside as long as we confine our questions to the here and now. On the other hand, it may having a bearing--not simply upon the "hereafter" or the "otherworldly," and not only upon religious faith or philosophy of a metaphysical sort--but even upon a fundamental enquiry into present realities. Indeed, the suggestion of eternity peeps out at us from the edges of our thought whenever we raise cosmological questions, such as: What was the state of the universe before this its present state? What was there in the beginning, aeons ago? The very word "aeon" is the Greek compound for, which "always is." To be sure, positivism declares an end to such questions; but its declarations seem to lack the kind of prohibitive force that is claimed for them. And so enquirers also go on to ask about the end and destiny of the universe: Does it have an end? When? What is its destiny? What would follow upon it? The European Enlightenment and the secularism that has grown out of it has attempted as a matter of principle and method to confine the interest of human intelligence to the finite and immanent circle of more or less immediate questions. One can at least wonder, however, whether scientific and philosophical enquiry can sustain itself if it is confined to such a circumscribed sphere of enquiry, as though it were to write "finis" at the very center of its effort to understand even before it has begun. And one can wonder further whether the civilization that must support such an enquiry into origins can maintain its own life without some understanding of that which--if anything--lies beyond the circle of finitude, and the circle of its immediate concerns. I should like to close, therefore, with a very brief look at time itself in order to see whether within its structures there may not be intimations of an order of duration that might include time and ground it. Or, at the very least, indications of how we might understand a little what such an order of duration would be.

If, in order to get an initial purchase on the notion of eternity, we look back within our own intellectual traditions, we come upon the famous text of Plato in the *Timaeus*.[23] There he sets forth the great unchanging forms and declares them to be immutable. Plato has often enough been interpreted as saying that the eternal forms are timeless; but his remarks elsewhere, especially in the *Sophist*, should warn us that for him the eternal is more than the mere absence of time, and that it encompasses life and power and existence itself.[24]

It is significant that Greek reflects the tension between measure and measured in two words it uses for time. *Chronos* stands for the familiar and everyday reckoning of time: the time marked by calendar and sun-dial in accordance with the movement of the heavenly bodies. Such chronology and chronometry directs attention to the succession of moments; and it is this that Plato clearly denies of the forms. But *kairos* stands for the fullness of time, as when we say that something is to be done "in due time," that is, "in its own good time," or "in its proper season"; or again, as when we say that "the time is ripe," and let time take its fullness from the thing of which it is the measure. This sense of time places the emphasis upon simultaneity. And it seems to me that it is the sense that is the source of an ambivalence in regard to the conception of eternity. For, the tradition to which Plato gives expression finds a certain ambivalence with regard to the meaning of eternity: it is in some sense timeless, yet not so that it is simply negative; for we can also say that it is in some sense timeful, yet not so that it proceeds from one moment to another in order to fulfill itself. In short, eternity is said to be both absence of time and fullness of time. St. Thomas Aquinas gives voice to this ambivalence when he insists that eternity includes all times, though--it must be added--by way of fullness (completeness, perfection) and not by way of a succession of moments.[25]

Now, a traditional way of arguing for the reality of that duration called "sempiternal" is to establish the existence of a fully perfect being which does not change, not because it cannot, but because it need not; and it need not because it possesses its being, its life, its knowledge and its freedom in the fullness of power. Such a being, then, has its own measure of duration, from age to age and forever, and what is more, as a complete and everlasting now *(nunc stans)*. Its stance, then, is not that of a static being (merely changeless), but the full presence of living power. This is the classical theological resolution by which time is rooted in eternity and issues from it, --an eternity, moreover, from which all order, including time itself derives.[26]

St. Augustine coupled the immutability of the eternal ideas with the very life of the Divine Being. And Boethius[28] defined eternity theologically as the simultaneously whole and perfect possession of endless life. The medieval Schoolmen distinguished several kinds of duration, since each kind of being was measured by that mode of duration appropriate to it: God by sempiternity (eternity in the fullest sense), angels and human souls by aeviternity (having a beginning but no end) and all other creatures by time (having both a beginning and an end).[29]. In a word: vary the mode of being, vary the measure; the measure is in accord with the way of being. For the Schoolmen, then, two factors comprised the meaning of "duration": it is a *measure*, and it is a measure *of being*. According to their view, time is not the only kind of measure. Rather, *qua* measure, time is the specific manner of duration that takes account of mobility; hence the emphasis upon succession in our ordinary understanding of time.

There is, however, a curious characteristic of the simultaneity of the phases of time that I have not yet remarked upon:[30] (i) The phases of time are, in the second-order sense of the term *simul*, simultaneous with each other in the *now* (simul^2); (ii) each phase is identically its correlative moment in its coming to be or passing away, so that it is the same past or present or future that is both phase and moment; consequently, (iii) there is a mediation through the phases of time by which the *moments* share a passing simultaneity with each other, a sort of transient contiguity in the order of duration.[31] This profound (though passing) simultaneity at the heart of time provides seed for reflection upon a duration that transcends change, a reflection prompted by Plato and the theologians. And if, in the words of St. Thomas, eternity "includes" all times, and includes them not merely potentially and indifferently (for that would amount to the mere exclusion of change) but in some sense actually, then a *new sense of process* is called for, --a process that retains the differences of time but without the partiality, without the relative emptiness and without the exclusion which is the negative condition for the dynamic succession that is the most obvious characteristic of time. Now, the Christian theologians already speak of processions within the immutable Godhead, --processions which are not successions in any ordinary sense of that term. They also speak of the unity of the divine differences, so that the inclusion of all times might be regarded from the human point of view as a *condensation* of actual duration, although it would in actuality be the very source of all forms of duration including time. Such a fullness may well remain beyond the reach of the philosopher, but there may be intimations of such a condensed duration (plenitude) in the permanence and simultaneity of temporal duration. Eternity, so understood, would be the highest unity of duration and the source of all temporal as well as trans-temporal order. It would be the ultimate and original implicate order.

Trinity College
University of Toronto

NOTES

¹The analysis which follows was arrived at by a metaphysical reflection upon the direct experience of time. In the larger sense of the term, then, it is empirical, though it is not based upon experimental results. Moreover, the term "implicate order" has been borrowed from David Bohm within the context of the present volume, because it raised the question of whether there are--if not different "implicate orders," --at least different ways in which a more general implicate order manifests itself. More particularly, in terms of the following analysis: Does time itself embody a sort of implicate order, largely tacit, and yet one that can be disclosed through metaphysical analysis? It may be that such an analysis will disclose properties different from those disclosed by experimental physics; but, since difference does not always mean contradiction or even incompatibility, the question remains whether they are in fatal conflict with or complementary to one another.

²Aristotle Physics 4.11. 2191: *arithmos kineseos kata to proteron kai husteron*. The translation is that of Richard Hope, *Aristotle's Physics* (Lincoln: U. of Nebraska, 1961) p. 80. In the *Loeb* edition (Aristotle IV: *Physics* 1-4. vol. 1, (London: Heinemann, rev. ed. 1957), p. 387), Wicksteed and Cornford render *arithmos* as "measure or dimension," adding (p. 388) that the number thus intended by Aristotle is the concrete numerable *(numeri numerati)* rather than the abstract numerator *(numeri numerantes)*. Aristotle also uses the term, *metron kineseos*, often interchangeably with *arithmos kineseos*, though sometimes with emphasis upon standard and precise quantification.

³Aristotle's discussion of the continuity of movement that lies at the basis of time *(Physics* 6. 1-4. 231^{a}21ff.) forestalls any identification of time with atoms, space or convention.

⁴In this respect calculation seems to differ from change in that it seems to be irrelevant to a column of figures whether its sum has been calculated before. One can argue, of course, that the subjective certainty of the correctness of the addition and the sum is not indifferent to the repeated recalculation. Nevertheless, this subjective certainty is to be distinguished from real change, since it measures our activity rather than any process among the numbers;

whereas in real change it seems that the objective reality brought about by movement is not in every sense indifferent to its repetition.

[5]Thus, in the Objective Logic *(Wissenschaft der Logik,* Book One), the categorial procession moves from *quality* to *quantity, quantum* and *ratio,* at each stage bringing along and sustaining the original qualitative base and eventually recovering the preponderance of quality over quantity in *measure.*

[6]See, for example, *Time and Free Will, Creative Evolution* and *Introduction to Metaphysics* throughout; also *Matter and Memory.* In an excellent chapter on Bergson in *A History of Modern European Philosophy* (Milwaukee: Bruce, 1954), especially pp. 818-819, 822ff. James Collins argues that Bergson's strictures against the quantification of time were more explicit than his positive characterization of the metaphysical nature of real duration. Bergson saw clearly that the intelligibility of duration could be accounted for neither through the abstraction of an essence, in Platonist fashion, nor by an *a priori* intuition, in Kantian fashion. But, Collins argues, Bergson failed to see that the intelligibility of real duration can best be accounted for through an existential judgment which recognizes time as embedded in existence. This would permit the valid employment of concepts in the representation of time. It is judgment alone, then, according to Collins, rather than intuition, that would have permitted Bergson to realize his intention of freeing metaphysical duration from the restriction to "lived" psychic time, and would have made it unnecessary for him to rest his case upon a trans-conceptual intuition and the figurative analogies he employs.

[7]*Physics* 4.13. 222^{a}10.

[8]*Physics* 4.11. 220^{a}3-10.

[9]For a discussion of this characteristic of an Aristotelian principle *(arche),* see my essay, "Analysis by Principles and Analysis by Elements," in *Graceful Reason: Essays in Ancient and Medieval Philosophy Presented to Joseph Owens, CSSR,* ed. Lloyd P. Gerson, (Toronto: Pontifical Institute of Medieval Studies): 315-330, esp. 322-323.

[10]Especially in *Ideen zu einer reinen Phanomenologie,* 1.3, c.2, secs. 81-82; ed. W. Biemel

(The Hague: Nijhoff, 1950); *Husserliana*, Bd. 3, pp. 199-201.

[11] John Wright has questioned whether the characteristics of time yielded by the analysis insofar as it goes beyond Aristotle's definition of change (as the measure of motion with respect to before and after) can still be attributed to real change, or whether the use I make of Husserl and am about to make of Kant does not restrict the analysis to time as it is experienced by human beings. Even more precisely, the question is whether the internal relations (inter-connections) between the moments of time do not presuppose memory and anticipation whereas much of non-human nature lacks them. In a word, I have humanized, and thereby subjectivized, time. The importance of the challenge cannot be exaggerated, since I do not mean the analysis to hold only for human subjectivity. I think that the idealism of Kant and even the idealistic elements in the later Husserl can be disengaged from some of the key conceptions thrown to the fore by their analyses. The issue puts the whole of one's metaphysics to the test, however, and to that extent cannot be answered adequately here. Nevertheless, let me say that the modes of perception, memory and anticipation do not *of themselves* intrude upon the truth and reality of what they disclose. Now, no truth about reality would arise from perception, memory or anticipation if the real world of objects were discontinuous atomic entities. The presuppositions of my own position, then, are: (i) that noetic media (perception, etc.) are not material instruments, but rather disclosive of their content, and (ii) that real development and real objects do have internal relations, including connection between what we discern as the stages of a continous development in time. This is, indeed, what is meant by the *continuity* of physical movement. Only if one assumes the discreteness of entities as fundamental (atomism, mechanism) does one have to face the problem of "putting them together" by means of subjective processes after the manner of Kantian critical idealism. To repeat, it is important to appreciate that I am not here talking about memory proper (recollection, recall, etc.), since it is an activity or process of human consciousness which presupposes a break in the immediacy of experience, a break that usually has not occurred in reality. That is, I recall something that happened yesterday, or somewhere else without recalling all that lies between here and now and then and there. Nor am I talking about imagination

which breaks down and recombines past components of experience in fictitious ways, since that also belongs to human consciousness alone and its order is not necessarily the order of the real world. When I talk about retention and protention, however, I speak of modes of consciousness that are immediate. Now, where they are involved in perception of the real, they provide the context for disclosing real change. Such a conscious context is not provided by physical things, of course, since they do not perceive, nor do they disclose the nature of the real world and its movement; they *are* that world. Nevertheless, these modes of disclosure have a foundation in reality, and that foundation is the continuity of movement. To be sure, the way in which things carry the past and future with them is not by immediate *memory* (retention) and immediate *anticipation* (protention) as such; rather, the continuity of movement guarantees a "carrying along" of a physical and not a conscious sort. The past is accumulated in residue (in physical things), in metamorphosis (in chemicals) and in growth (in organisms); the future is implicated in the present stage of a development by the inertia of direction, impetus and inclination (in physical things), by affinity (in chemicals), and by tendency and appetite (in organisms and animals). Retention and protention are the faithful context within which the interrelated continuity of movement shows itself.

[12]Ibid., p. 199.

[13]"The measure of real movement": As measure, time does not change; but as measure of change, it is more than a merely formal pattern imposed indifferently upon whatever happens to change. If it were merely formal, if the moments were merely empty, then its measure would be arbitrary. Yet, on the other hand, if it changed, it would not measure change. Aristotle remarked that time does not measure movement with regard to more or less without qualification, but with regard to fast and slow, just because it is the measure of the magnitude of a movement over distance in regard to "more or less" where the magnitude has the value of successive units of duration. (See *Physics* 4.12 220^b1ff.) It is here too that Aristotle distinguished between "the counter that counts" and "the dimension that is counted which [time itself] really is." To further confirm that time is not simply movement, however, Aristotle pointed out that time is also a "measure of rest, for all rest is in time." (ibid.,

2211b7ff.) Once again, it is important to recognize that we are not here considering any definite time-scale, --since that can be relative either to different scales of motion or velocity or to different human intentions and attentions or to a combination of man and nature, --but rather we are considering time in its qualitatively differentiated structure.

[14] A consideration that followed by a century or more the earlier scientific and philosophical reflections upon the nature and role of space in modern enquiry.

[15] *Critique of Pure Reason*, B219 A177: "Die drei modi der Zeit sind *Beharrlichkeit, Folge* und *Zugleichsein.*" N. Kemp Smith translates *Beharrlichkeit* as "duration," but later (B224 A182) as "permanence." The latter rendering is preferable. If some form of *duro* is chosen to translate *Beharrlichkeit*, however, a clearer rendering would be "endurance." Either "permanence" or "endurance" will reserve the term "duration" for the generic character of time, including therefore both its modes and its moments.

[16] Cf. Aristotle, *Physics* 4.10. 218a4ff.

[17] See especially Book 9. Even more precisely, time does not resemble the *productive* active potencies, since time does not produce anything; it only measures production in a certain way. Its closest analogue to concepts of power is to that of immanent activities, such as knowing, choosing and loving. This suggests that time may be of a spiritual order and have a high degree of intrinsic self-constitution.

[18] This consideration is sufficient to satisfy the demands of time as the measure of motion, and is not open to the third man argument, viz., that second-order permanence needs a further permanence and so on *ad infinitum*. Nevertheless, we have reached a point at which eternity (eternal duration) suggests itself as the ultimate ground of time, though a further and difficult analysis would be required to make that claim.

[19] I am apeaking here of the *fundamental qualitative structure* of time. I do not, therefore, find myself in contradiction with the conception of the relativity of time, since that bears principally upon the quantification of the time-structure and the qualitative properties that follow therefrom.

[20] In his examples Kant tended to blur the distinction between numerical order and the counting which presupposes time, but it seems to me that his intention was to maintain the distinction.

[21] Is there, then, a sense in which time is accumulative? Does it, in a fashion and so to speak, "gain weight?" We do speak of the "weight of years," but that seems to be a reference to either individual aging or to social tradition, i.e., reference to change of a definite type and content.

[22] It is intriguing to consider that time may be the very presence of spirit to movement, but that would require further enquiry. Cf. fn. 17 above.

[23] 37e-38a: "For there were no days and nights and months and years before the heaven was created, but when he [the maker] constructed the heaven he created them also. They are all parts of time, and the past and future are created species of time, which we unconsciously but wrongly transfer to eternal being, for we say that it 'was,' or 'is,' or 'will be,' but the truth is that 'is' alone is properly attributed to it, and that 'was' and 'will be' are only to be spoken of becoming in time, for they are motions, but that which is immovably the same forever cannot become older or younger by time, nor can it be said that it came into being in the past, or has come into being now, or will come into being in the future, nor is it subject at all to any of those states which affect moving and sensible things and of which generation is the cause. These are the forms of time, which imitates eternity and revolves according to a law of number." The translation is by Jowett in *Plato· The Collected Dialogues*, ed. Edith Hamilton and Huntington Cairns (New York: Bollingen, 1963), p. 1167. I might add that the choice of Jowett's translation is deliberate, just because it reflects that reading of Plato that was ingredient in the theological tradition.

[24] The Stranger asks Theaetetus (249a): "But tell me, in heaven's name, are we really to be so easily convinced that change, life, soul, understanding have no place in that which is perfectly real--that it has neither life nor thought, but stands immutable in solemn aloofness, devoid of intelligence?" (Op. cit., p. 993)

²⁵*S.T.*, I, a.10, a.2 ad 4: "Words denoting different times are applied to God because His eternity includes all times, and not as if He Himself were altered through present, past and future." The last is added to ensure that the "inclusion" is not a simple containment in which the different times are present in eternity in their transience. In a word, they are there; they are not simply absent from eternity. But they are not there as they are in time itself. They are there in a higher way *(eminentior)*.

²⁶So that the theological argument resolves time by grounding it in the eternal being. Other attempts to ground time take anthropological and cosmological forms. The difficulties of these attempts are the reduction of time to human subjectivity, on the one hand, and to physical change, on the other.

²⁷*De diversis quaestionibus*, q. 46.

²⁸*De consolatione philosophiae*, Book 5, Prose 6.

²⁹See St. Thomas Aquinas' classical discussion in *Summa theologiae*, 1a a.10, especially a.5.

³⁰For an earlier and less detailed analysis of the character of simultaneity, see "A Moment of Truth: Present Actuality," in *The Review of Metaphysics*, 33/4 (1980): 680-684. The important distinction between *phase* and *moment* is not clearly made in that article, whose focus is more upon metaphysics than upon time.

³¹It is this feature upon which classical metaphysicians insist when they distinguish a *per se* (hierarchical, essential or "vertical") series of causes from a *per accidens* (successive, accidental or "horizontal") series of causes. Even Kant struggled to articulate this need for contact, though he diminished the sense of communication in causality to a vanishing point. In his *Critique of Pure Reason*, he writes (A203 B248): "If I view as a cause a ball which impresses a hollow as it lies on a stuffed cushion, the cause is simultaneous with the effect." He adds, however, that a difference in time (with respect to the state of the cushion) is the only way in which we can empirically determine the difference between a particular cause and a particular effect. And so, instead of accepting the classical metaphysical conception of cause as the communication of something to another, he defined it as an event that necessarily precedes another event which is its effect. Nevertheless, he never quite managed to

eradicate entirely a certain "communication," and only an abandonment of the category of causality can appear to eradicate the sense of a contiguity that is more than a mere adjacency. But that discussion lies beyond our present analysis.

COMMENTS ON THE PAPERS

David Bohm

Firstly, let me say that the papers given in this conference struck me as being on a very high level. They make a number of valuable criticisms of and comments on my notions of the implicate order, and introduce a great many interesting and significant ideas of a theological nature, which are related to these notions. They also bring back to me many enjoyable memories of the conference itself and of the lively exchange which took place between all of us at that time. So I am glad to have this opportunity here to make some further comments relating to these papers.

First, I shall consider Frederick Crosson's paper, "Man and Meaning of the Whole." Crosson begins by pointing out what he feels to be some unclarities in the basic concepts of enfolding and unfolding. Does enfoldment have to do primarily with information about the whole enfolded in each part (as in the holograph)? Or is what is explicit in the whole implicit in the part (as in the meaning of a thought)? Or is what is unfolded a transient or non-independent part of a whole in which it is embedded and from which it only *seems* to emerge as a separate entity (as in my comments on Polanyi's discussion of bicycle riding)?

In answer, I would first say that all three of the strands described above are contained in the notion of the implicate order, and that they are woven together so as to constitute a whole. With regard to information, I would emphasize that this is not merely passive, but has, in general, an active and creative role[1] in determining entities as *internally* related to each other. An evident example is the set of possible forms of the DNA molecule of a species which constitutes a common pool of information that guides the pattern of growth and development of each individual member of the species. In certain ways, this information enfolds the entire past experience of the species. But more important, the possibility of "carrying out" what is "meant" or implied by the information is enfolded in the material structure of the cell and of the environment as a whole. Here, I would refer also to the example of the seed of a plant, which supplies very little of the material or energy required to create the plant. This latter arises from the over-all environment, which is "informed" by the seed to unfold a plant, instead of just exlicating itself more or less constantly as inanimate matter.

The basic point that I want to make here is that the implicate order is a principle that works throughout every phase of existence, including the mental and the physical, information and the content that gives it meaning, and the ground (e.g., of the holomovement) along with that which emerges from the ground.

Crosson then points out that analogies to the implicate order, such as those that I have given, are necessarily only partial, and asks whether the implicate order has any real *instances* beyond that of quantum wholeness. To this, I would say that the prime instance of the implicate order is consciousness itself. Thus, the consciousness of an individual human being enfolds the entire world of his personal experience, of his society and culture, as well as of his past history, going on to the history of his society and of the species, and beyond ultimately to that of the entire development of the universe. Such enfoldment is active, in the sense that a person's whole response, his mode of being, is fundamentally affected by all this.

Here, it seems that the idea of separate substances suggested by Crosson as a counter-example to the notion of mutual interdependence is at variance with most of what is known about the nature of reality. Thus, not even the elementary particles of physics, with their creation out of the vacuum and annihilation back into the vacuum, along with their apparently unlimited transformability, can properly be thought of as separate substances. At the other end of the scale, the human body, for example, is certainly not a separate substance, as its matter comes from the environment and is constantly exchanged with the environment. And with regard to the mind, it seems clear that starting from infancy, this is fundamentally affected by society and by culture. Each one of us *is* the result of all this (though he is also more, and this of course is what makes freedom possible).

In connection with Crosson's statements about machines, I would refer to what Cobb said at this conference. A machine is admittedly more than the sum of its parts, but the parts are little affected by their role in the machine.

I would certainly agree with Crosson in his emphasis on Polanyi's notion that the particulars of an entity may be governed by the organizing principles of a higher entity that they form. This is indeed also essential for the implicate order. But I would add that in a machine, these higher organizing principles can work only through affecting the boundary conditions on the fixed mechanical laws governing the particulars. In the implicate order, however, the constitution of the particulars and even the very laws that apply to them ultimately unfold from the whole. This unfoldment will be by way of the organizing principles that have been mentioned, which operate in more subtle and more comprehensive levels. This leaves room for a much more dynamic and creative penetration of the mundane by the transcendent than would be possible if the former were governed by nothing but mechanical laws. So, while I can only agree wholeheartedly with the notion that a higher level of meaning organizes the lower levels, I would simply add that it seems unnecessarily restrictive to suppose that this can happen only through the imposition of boundary conditions on the mechanical laws of these lower levels. (A notion similar to my own as described above appears at the end of Hill's paper, where he points out that the implicate order may refer to an intermediate world between the transcendent and the level of ordinary explicate experience).

This brings me to John Cobb's paper, "Bohm's Contribution to Faith in Our Time." On the whole, I find it hard to add very much to what Cobb says, as the two of us are in agreement on a wide range of fundamentals. I appreciate that one of the things that Cobb wants to do in this paper is to try to correct what he feels to be a tendency on my part to over-emphasize the whole at the expense of the parts. It is clear that there is danger in overstating the case in either direction. And for this reason, I think that it is of value for Cobb to draw attention to this danger, as he has done.

In order to bring out the two polar sides of the process, involving whole and part, I would emphasize that there cannot ever be a question of capturing *the* whole (i.e., the totality) in the context of any thought. What we can do is to be careful to give due weight to the *quality of wholeness* This includes the notion that all *is* undivided flux or movement, without any real break of seam, and that in this movement, there are to be found relatively independent and lasting sub-wholes that are internally related. Each sub-whole participates in a greater whole, and so on. The whole and the sub-wholes are thus also not divided from each other.

If we call the sub-wholes parts (as I have stated in my paper for the conference), one can bring out the meaning of the quality of wholeness by considering two alternative principles:

The wholeness of the whole and the parts

The partiality of the parts and the whole

In certain contexts, we use the former of these principles, and in other contexts, we use the latter. Thus for the sake of analysis, we may, at least in our thoughts, distinguish a society from its members, and treat the whole in terms of society *and* its members. But to regard this approach as the primary one and to give it major emphasis in all knowledge is equivalent to adopting a mechanistic philosophy. What is being suggested here is that the primary emphasis has generally to go to the wholeness of the whole and the parts, and that to do this is to go beyond mechanism.

Cobb also raises questions with regard to my use of language about time. As brought out in Schmitz's paper, time is a very subtle notion, and difficult to talk about coherently. In this regard, the relationship of eternity and time is particularly difficult to discuss. I shall return to this question in my comments on the papers that follow. Here, I shall only say that the very notion of some universal features of time as a whole (e.g., that each moment is, as Whitehead says, uniquely creative in some ways) seems, at first sight at least, to presuppose that one is speaking from a kind of eternal standpoint, i.e., one from which a vast range of moments (or actual occasions) are "seen" to have the qualities in question as if being "viewed" all at once." I would say that eternity is not such a whole (or totality) of time, but rather, that it is the *quality* of the wholeness of time. According to the principle of the undivided wholeness of the whole and the parts, the eternity of eternity and time has thus to be the primary or major principle, even though at suitable levels of abstraction, eternity and time can be put side by side into some kind of relationship, as appears to be the case in Whitehead's philosophy. If one asks what is the relationship of eternity and time considered as such partial categories, one is therefore already really adopting a standpoint beyond both. It is this which I find is closer to a true sense of the meaning of eternity than is the mere notion of "the whole of time."

I think that almost all would agree that if this sort of notion of eternity has a real meaning, it must in some ways be paradoxical. Perhaps this paradox can eventually be resolved, but it seems more likely that eternity is something that cannot be explained by thought, which latter is, in certain essential ways, bound to time. Perhaps one could say in this connection that eternity unfolds and manifests itself in time through the intermediary of higher organizing principles that determine the order and measure of time at lower levels. I have discussed this point elsewhere[2] in more detail but wish to add here only that the theory of relativity shows that the order and measure of time are not given a priori but are dependent on the gravitational potentials (which in turn depend ultimately on the state of the whole universe). This leads to very subtle and difficult problems at the moment of the supposed origin of the universe as a "big bang," in which none of these potentials can be defined. Does this not suggest that we are approaching a point where some order beyond that of time is relevant, and that this may well be the source of the organizing principles governing the eventual enfoldment of our commonly experienced order of time? But I must defer further discussion of this point until I comment on the papers of Wright and Schmitz.

I now come to William Hill's paper, "The Implicate World: God's Oneness with Mankind as a Mediated Immediacy." This paper raises many very deep and subtle questions, and I can comment on only a few of these here. As Hill points out, science (at least as this is now commonly defined) cannot address itself directly to the deepest dimensions of human reality. But I would add here that modern science, with its instrumentalist and positivist bias, tends at least tacitly to devalue these dimensions, by denying them any reality except in some abstract, insubstantial and therefore implicitly unimportant kind of mental or spiritual domain. What I feel is that the implicate order leaves room for the possibility that these dimensions can operate, even in the mundane world. Even more, as I have already suggested earlier, it fits in naturally with the notion that the whole ultimately originates in and is sustained by a transcendent ground. This ground is not a mere *concept* of being. Rather, the very act of knowing (and loving) is, as I understand Hill's thesis, an act of being, which is, in some analogical sense, also a grasp of the quality of being itself. Such *esse* is the act giving rise to essences, which differentiate the species and issue forth as variety. So, knowing is an analogy to being, but also a being which, in its essential act, produces this very analogy.

Thus, in some sense, knowing is able by analogy to throw light on itself. This is possible because, as Hill says, "the human intellect ... grasps the analogical character of being, i.e., the relative or proportional unity of all realities ... in an intentional act that is at once conceptual and judgmental." There is, in this way, "a non-conceptual aspect to intellection which is always at the same time conceptual." Or as I would like to put it: *what* we think can be expressed in terms of concepts, but *how* we think is tacit, an act that is ultimately not definable in terms of the content of this thinking.

I feel that all of this is compatible with the implicate order when this is viewed in its broadest and most fundamental sense, which must include the deeper organizing principles emerging ultimately out of a ground (implied by the holomovement) that is beyond the distinctions of space and time and that is not definable in any way at all.

What is particularly important here is the notion of "isness" as analogical. This allows us not merely to reflect conceptually on being, but to participate deeply in the latter, even in our very act of reflection. As shown in my book,(3) this sort of notion of analogy can very naturally be assimilated into the implicate order, through the idea of order as similar differences and different similarities. These are essentially put as *qualitative ratios* such as: A is to B as C is to D. Such ratios are evidently ideally suited to the expression of an analogical relationship.

It would be difficult for me to comment here fully on all the subtle content of Hill's paper. I have found particularly interesting what he says about the relationship of God and the created world. In a way, he seems to be proposing that God is not conditioned by his knowing this world, nor by past "experiences" with it, though he is nevertheless affected by it in a creative way. Within the created world, however, there is not such absolute freedom. Rather, there is the possibility of freedom within a context given by nature (including human nature) and by history. I would add here that one of the most serious limitations on this freedom is man's tending to become conditioned to a generally false mode of operation of mind and heart. (Recall here that the basic meaning of "false" is "deceptive," rather than merely "incorrect.") One of the difficulties that I do not think has been satisfactorily resolved is to understand why and how this possibility of self-deceptive falseness (or more accurately false-heartedness) has been implanted in mankind in the absolutely free act of a loving God. This God has created human beings as capable of being free within their appropriate context, and yet capable of being infected from earliest childhood with such a "disease" endemic in this context, that entrains an almost total loss of real freedom.

The notion of parentheism outlined in Hill's paper is close to a way that I have developed for putting the implicate order.(2) There, I have proposed that in some sense, the temporal succession that is basic to the existence of such finite sub-whole is in some way contained (or perhaps enfolded) in eternity. The relationships within time are considered to be "horizontal," while the relationship of eternity and time is, in some sense "vertical." The "vertical" is however ultimately implicit and not definable. And the horizontal relationships ultimately unfold from the vertical, so that it is through the apprehension of these that the vertical is, in some sense, implied or revealed. I feel that this approach is close to the "mediated immediacy" discussed in Hill's paper (i.e., that the horizontal mediates the vertical, which is, however, the true immediate). In this connection, I would agree with Hill, that eternity is a standpoint outside of time rather than something that contains a fore-knowledge or predetermination of the whole of time. That is to say, eternity contains time, and yet, it does not *determine* the future completely in terms of the past, nor in terms of something given statically that rules over all time. (The notion of the totality of space-time as a single block is thus not equivalent to that of eternity.)

As I stated earlier, these are very difficult questions that probably will never be entirely free of paradox, but I feel nevertheless that they are important, in the sense of touching on something that is at the heart of our being, and indeed, of all being.

The next paper is John Wright's "Cosmic and Human Evolution." This deals with the latest theories of modern science with regard to the evolution of the universe and of humanity. His whole approach is centered on the insight of Thomas Aquinas that the order of things to one another is on account of their common order or end. The grasp of such order is thus essential to grasping or understanding the nature of reality as a whole.

If we go into current theories of development of the cosmos as we know it, we start from the presumed origin of the universe in an initial "big bang," in which conditions are such that nothing that is now familiar to us can exist in any well defined form, neither large-scale bodies--nor atoms, nor electrons and protons, nor quarks, nor any other kinds of particles. Indeed, as I have already indicated in my comments on Cobb's paper, not even the gravitational potentials can be defined. From this it follows that the measure of space-time had at this point no meaning. Nor was it possible to distinguish past from future, here from elsewhere, and the like.

As Wright points out, within this theoretical framework there does not seem to be much meaning to considering a time before the universe began, nor is there anything outside the universe, into which it could expand. Nevertheless, somehow this initial uncharacterizable point of concentration of pure energy seems to have exploded and started to radiate outward at the speed of light, to give rise to space, time and matter. In the beginning, there were no particles, but as the universe cooled down, particles formed and became stable. Later, combinations of particles became stable, and these then formed organic self-replacing structures, moving on eventually to human beings with their self-conscious unifying centers.

All of this fits very well, of course, within the notion that I have discussed in the previous comments that the implicate order provides an intelligible mediation between the transcendent and the mundane. For example, as I have suggested earlier,[3] according to modern physics empty space contains a zero point energy which is immensely beyond anything thus far known to us (e.g., in one cubic center of empty space is an energy far greater than would be liberated by the annihilation of all the matter in the entire universe). Matter as we know it is a small relatively stable and independently moving "ripple" on this vast "sea" of energy. We, who are constituted physically out of such "ripples," do not perceive this "sea" (any more than would presumably a fish who is swimming in the terrestrial ocean be aware of this ocean). What I proposed was that as a fortuitous combination of water waves sometimes produces a gigantic wave that appears suddenly at a certain place, so some combination of factors may have produced a highly localized point of excitation of the "vacuum." This then expanded to give our universe of space, time, and matter. But as I have explained elsewhere,(2) this "sea" is not to be understood in terms of ordinary notions of space-time.

Rather, we should regard this as a kind of "relative eternity,"(2) or eternity that is in some sense alive and in movement, but not in the order of time and space, as we ordinarily understand it. This "sea" of energy, that contains inplicit organizing principles similar to those discussed in my comments on the papers of Crosson and Cobb, enfolds a process from which emerges our universe of space, time, and matter (and perhaps other universes as well).

Along the lines indicated by Wright, one can see that what is enfolded in the original "sea" of unformed energy also contains the possibility of human beings. Mankind is thus "propelled" by the whole of its past, going on to the very origin of the universe and into the depths of the primal "sea" of unformed energy (or beyond). Perhaps this corresponds to the Alpha of Teilhard de Chardin. Part of the activity of mankind is a response to the surrounding context of the space-time continuum. But another part is a response to a deeper destiny or "purpose" that is enfolded still more subtly and inwardly in the consciousness of man and ultimately in the transcendent ground of reality that goes beyond what can be defined in the context of any thought. This would be the Omega of Teihard de Chardin, which grounds the relationships of all things to each other, and which is in turn furthered by all these relationships.

I think that all of this constitutes a reasonable way of looking at the cosmology implied by modern science. But in order further to clarify what this cosmology means, I would like to ask one question here:

why is it necessary to go through this whole process? In some sense, for example, mystics are saying that the end can be reached immediately, without the need for further evolution. Or alternatively, as Hegel suggested, this very process, is, at the same time, both end and means, so that the end is both eternally realizing itself and externally realized. This suggests to me that perhaps the meaning of this whole process of evolution can be properly understood only from a standpoint beyond time, along lines discussed in Hill's paper.

The above question leads naturally to the final paper to be commented on here, which is Kenneth Schmitz's "Temporal Integrity and the Implicate Order." This paper contains a penetrating analysis of the concept of time, which has a significant bearing on the meaning of this concept in the implicate order.

The first important point made in this paper is that the depiction of time or a line as if it were a kind of space consisting of co-existent points is an analogy of only limited value. For time also implies a kind of negation in the sense that the actuality of a given moment is the non-actuality (or non-existence) of all others. And the alternative idea of time as a point moving along a line implies yet another kind of time in which this point moves, a kind not represented by the line at all. The problem of the nature of time is in this way merely shunted to a new level. Another notion of time is that it is just a feature of movement. But this leaves out the *enumeration* of successive intervals which, as Aristotle points out, are an essential constituent of the concept of time. That is to say, though similar events may recur in a movement such as that of the tides, there is a basic difference in that *today's* tide does not return tomorrow, no matter how similar tomorrow's tide is to that of today. For tomorrow's tide will have a past which includes today's tide, and thus it will have a past that today's tide does not have. This structure, resembling that of Chinese boxes within boxes, is characteristic of how the irreversibility of time is contained in the implicate order, in which each moment enfolds its past, in a way in which it does not enfold its future.[2] The enumeration of time thus corresponds to the enumeration of successive stages of enfoldment.

But as Schmitz emphasizes, time is also not simply enumeration. Thus, it is also the *measure* of movement.

In the first instance, this measure has a strongly conventional element (e.g., in the choice of seconds, minutes, hours, and the like as units of time). But ultimately, such measure cannot be entirely conventional. There are natural units of duration in each movement, as for example, the seasons measure the growth of plants. Modern physics has indeed found in the periods of atomic vibrations a natural measure that holds universally.

But even measure is still not the most fundamental feature of time. Rather, it has in addition the basic qualitative feature of "before and after," which gives rise to the order of time that I have already mentioned as corresponding to the order of enfoldment in the implicate order. That is to say, time measures movement with respect to an order unfolded from a natural succession of "befores" contained in "afters," which in turn play the roles of "befores" in yet later "afters."

The succession of moments not only exhibits differences, it also exhibits inter-connectedness. Thus Aristotle has remarked that the present holds past and future together, and is their common boundary. The present is thus a constitutive principle of time, since it both differentiates past and future and connects them. To approach time in this way is clearly to emphasize the actual, and the basic principle of time works in terms of the distinction of actuality of the present and the non-actuality of past and future.

In terms of the implicate order, I would put this notion as the actuality of the present, which actually enfolds past and future, but in different ways (which will be discussed shortly). Though the enfoldment of, for example, the past is actual, the past itself is no longer actual as such, but is rather actually a mere form lying on (or in) the present.

Schmitz points out however, that this essentially Aristotelian analysis of time can be greatly improved by considering the more recent contributions of Kant and Husserl. This requires a recognition of three "dimensions" of time, which are the *moments* of time, the *phases* of time, and the *features* of time. Let us begin with the present moment, which is actual. This *now* is a real duration, or rather, a sort of triad of duration. For in it that which is just-passing-away is still conjoined with the actual present, even while that which is not-quite-yet is also conjoined with the same present. Or to put it differently, there is an unbroken *retention* of the past in the *now* and *protention* of the future in the same *now*. Such retention is not the same as recollection, which latter is the reconstitution of an image of the past from some kind of record stored in the depository of memory. Similarly, ordinary anticipation is also a kind of image constituted out of the projection of the implications of past knowledge into the future. Retention and protention are not thus constituted out of a record, or out of knowledge, but are present and actual, an intrinsic part of the actuality of the present moment. That past which is gone and held only in memory belongs to a different *moment* to be distinguished from *now* as holds also for that future which is merely anticipated. That past which is *now* experienced as retention and that future which is *now* experienced as protention will then be called *phases* of the moment. Each moment thus contains three phases, which are *present*, *past* and *future* all of which are also *now*.

A closely related notion is also basic to the implicate order.(2,3) Consider for example, the experience of listening to music. At the present moment, a certain note is perhaps being sounded. But there is a retention in awareness of a number of the previous notes, which is not a mere recollection in memory. Rather, there is a sense of "reverberation" of the notes, not only as an actually felt experience of sound that is fading in intensity and steadily becoming non-actual, but also as a series of responses to the music, which are emotional, intellectual, physical, and the like. Similarly, we are even now actually disposed toward a certain range of possible future developments from these notes, and we may be shocked if the actual development is too far from that toward which we are disposed. This may evidently be regarded as a form of protention.

I have suggested(2,3) that such retention and protention are a kind of co-present sequence of enfoldments, which rise gradually to actuality and then gradually fall below the level of awareness to become non-actual as longer intervals of time come into play. The sense of continuity of the music evidently depends on the co-presence of all these phases. Thus, for example, if notes are played several seconds apart, we will no longer experience the unity of the theme. So I have proposed that in such co-presence of phases, we directly experience the implicate order, and that indeed, this is much more direct than is the experience of the explicate order, which depends on previous familiarization with the implications of quite abstract kinds of thought.

The distinction of past and future is then this: The past, held in retention is continually passing from actuality toward non-actuality, while the future, held in protention, is constantly passing from non-actuality toward actuality. When this notion is extended, we may say also that past *moments* have already passed from even the degree of actuality characteristic of retention while future *moments* have not yet reached the degree of actuality characteristic of protention. It is through this analysis in terms of phases (which I regard as stages of enfoldment) that one is able to understand the continuity of time. What is crucial here is that the distinction of actual and non-actual is not an absolute dichotomy, but proceeds through a sequence or succession of *degrees of actuality* (which I treat with the aid of the notion of degrees of enfoldment).

This brings us to the consideration of the "modes" or "features" of time, which Schmitz has developed on the basis of notions found in Kant's work. These are *permanence succession* and *simultaneity*.

Firstly, with regard to succession, we can say that this is an internal relationship in a sequence of *nows*. For each now contains the others, through retention and protention, as well as through memory and anticipation. I would add that this internal relationship is one of enfoldment, and is thus characteristic of the implicate order.

We then come to the feature of permanence. What is permanent is that which abides, which remains throughout a change of state. But most deeply, what abides is not what is subsistent or "residual" in things (e.g., the record of memory). For ultimately, this too will pass away. Rather, it is the *activity and dynamism of the entire time structure itself* as it has been described here in terms of moments, phases, and features. Though everything changes, there is a sense in which the basic structure of time is what nevertheless abides in all the dynamism.

To this, however, I would raise a question. If one accepts the "big bang" hypothesis, can one not say that even the basic structure of time is in some way limited along the lines that I have discussed in earlier comments? Or would one bring in the assumption of yet another level of time that could, for example, be valid in the immense "sea" of energy out of which the universe as we know it may have emerged from a small region of strong disturbance?

We then come to simultaneity. Simultaneity means co-existence. But this notion requires additional specification. Schmitz remarks that the features--permanence, succession and simultaneity--are themselves simultaneous with each other. There is also a further kind of simultaneity, which is the co-presence of the three *phases* of time to one another. These kinds of simultaneity are not merely the co-presence of *things* in a particular moment. Rather, they are aspects of a permanent co-presence of the dynamic features of phases of time itself.

The notion of co-existence, however, requires yet further explanation. Things that co-exist may act on each other, at least in some ways, whereas things that are no longer actual or are not yet actual cannot act in this way. However, the theory of relativity does not allow instantaneous control of spatially separated moments, but restricts contacts to those transmitted by the speed of light or less. This implies[2] that, ultimately, moments cannot be extended in space nor can they have duration of time. Rather, they must be dimensionless "point events." Such a notion is not compatible with the structure of time discussed here, which seems to be necessary for providing an intelligible basis to the continuity of time. Therefore, I have questioned[2,4] whether the theory of relativity in its present form can be entirely adequate, and have suggested that it may be an approximation to a deeper theory, in which the notion of moments with extension and duration would play an essential part.

Finally, Schmitz raises the question as to where there is present in time the seeds for an understanding of eternity. Here, he refers to Plato, St. Augustine and Boethius, who argued that eternity is not just the absence of time (and even less, of course, just the totality of time), but rather that it encompases life, power, and existence itself. Also, he points out that the Greek *Chronos* stands for the familiar reckoning of time, while *kairos* stands for the fullness of time. When for example we say "the time is ripe" or that something happens "in its own good time" we are giving voice to the latter meaning, which implies that time emerges from the enfoldment of essential being, and is thus not a mere measure of process.

If the above notion is extended without limit, we come naturally to the idea of eternity as enfolding and unfolding time, which we have discussed earlier. Each moment is then the unfoldment of eternity, and the moments differ primarily in their difference of unfolded content. As we go deeper into the enfoldment order of each moment, we come to a greater and greater similarity and internal relationship of moments. As I have brought out elsewhere,[2] it may be that in this way all moments are ultimately one, and that this is eternity. This means that eternity is the life, power, and existence that has been mentioned above, and that this is not something far away, but rather the most immediate ground of the being of everything. The differences of time, space, quality, and the like would then mediate what is the truly and ultimately immediate which is universal, and which contains all such differences enfolded within it. But, of course, as in the comments on Hill's paper on mediated immediacy, this does not imply a determination of future in terms of past or of all time. Rather, as pointed out earlier, the whole process is creative, in ways that are perhaps not capable of being expressed without some kind of paradox. This seems also to be in line with what Schmitz is implying in the concluding section of this paper.

Birkbeck College
University of London

REFERENCES

(1) This issue is discussed in D. Bohm and B.J. Hiley, *Foundations of Physics* to be published, and also in D. Bohn, *Journal of American Society for Psychical Research* to be published.

(2) D. Bohm, Claremont Conference (unpublished).

(3) D. Bohm, *Wholeness and the Implicate Order* (Routledge and Kegal Paul: London, 1980), esp. Chaps. 5 and 6.

(4) D. Bohm and B. J. Hiley, *Found. Phys.* 14 (1984), p. 255.

HIDDEN VARIABLES AND THE IMPLICATE ORDER*

David Bohm

I have been asked to explain how my ideas of hidden variables tie up with those on the implicate order, and to bring out in some detail how both these two notions are related. In doing this, it would perhaps be best to begin with an account of how I came to these ideas in the first place.

The whole development starts in Princeton around 1950, when I had just finished my book *Quantum Theory*. I had in fact written it from what I regarded as Niels Bohr's point of view, based on the principle of complementarity. Indeed, I had taught a course on the quantum theory for three years and written the book primarily in order to try to obtain a better understanding of the whole subject, and especially of Bohr's very deep and subtle treatment of it. However, after the work was finished, I looked back over what I had done, and still felt somewhat dissatisfied.

What I felt to be especially unsatisfactory was the fact that the quantum theory had no place in it for an ad equate notion of an independent actuality--i.e., of an actual movement or activity by which one physical state could pass over into another one. My main difficulty was not that the wave function was interpreted only in terms of probabilities, so that the theory was not deterministic. Rather, it was that it could only speak in terms of the results of an experiment or an observation, which has to be treated as a set of *phenomena* that are ultimately not further analyzable or explainable in any terms at all. So, the theory could not go beyond the phenomena or appearances. And basically, these phenomena were very limited in nature, consisting, for example, of events by which the state of a particle could be ascertained. From a knowledge of this state, we could go to a wave function that predicted the probability of the next set of phenomena, and so on.

On thinking what all this meant, it began to occur to me that the quantum theory might actally be giving a

---

*Previously published in *Zygon Journal of Religion & Science* 20/2 (1985):111-124.

fragmentary view of reality. A wave function seemed to capture only certain aspects of what happens in a statistical ensemble of similar measurements, each of which is in essence only a single element in a greater context of over-all process. Although von Neumann had given what purported to be a proof, that to go any further would not be compatible with the quantum theory (which was already very well confirmed indeed), I still realized that mathematical proofs are based on axioms and presuppositions whose meanings are often obscure, and always in principle open to question. Moreover, the theory of relativity, which was also regarded as fundamental, demanded a space-time process (e.g., one that could be understood in terms of fields) which constituted an independent actuality, with a continuous and determinate connection between all its parts. Such a process could not be treated solely as a set of fragmentary phenomena that are statistically related.

This requirement becomes especially urgent when relativity is extended to include cosmology. It seems impossible even to contemplate the universe as a whole, through a view which can discuss only in terms of discrete or distinct sets of phenomena. For in a cosmological view, the observing instruments, and indeed, the physicists who construct and operate them, have to be regarded at least in principle as parts of the totality. There does not seem to be much sense in saying that all these are nothing more than organized sets of appearances. To whom or to what would they appear, and of what would they be the appearances?

I felt particularly dissatisfied with the implicitly schizophrenic attitude of accepting the independent existence of the cosmos while one was doing relativity, and, at the same time, denying it while one was doing the quantum theory, even though both theories were regarded as fundamental. I did not see how an adequate way of dealing with this could be developed on the basis of Niels Bohr's point of view. So I began to ask myself whether another approach might not be possible.

In my first attempt to do this, I considered a quantum mechanical wave function representing, for example, an electron, and supposed that this was scattered by an atom. By solving Schroedinger's equation for the wave function, one shows that the scattered wave will spread out more or less spherically. Nevertheless, a detector will detect an electron in some small region of space, while the extended spherical wave gives only the probability that it will be found in any such region. The idea then occurred to me that perhaps there is a second wave coming in toward the place where the electron is found, and that the mathematical calculus of the quantum theory gives a statistical relationship between outgoing and incoming waves.

However, to think this way requires that we enrich our concepts to include an incoming wave, as well as an outgoing wave. Indeed, since further measurements can be made on the electron, it follows that the second wave spreads out, it may give way to a third, and so on. In this way, it becomes possible to have an on-going process in which the electron is understood as an independent actuality (which will, of course, give rise to phenomena through which it may be detected). One is thus implying that the current quantum theory deals only with a fragmentary aspect of this whole process -- i.e., that aspect which is associated with a single observational event.

It seems clear that at this stage, I was anticipating what later became the implicate order. Indeed, one could say that ingoing and outgoing waves are enfolding and unfolding movements. However, I did not pursue this idea further at the time. What happened was that I had meanwhile sent copies of my book to Einstein, Bohr, Pauli and a few other physicists. I received no reply from Bohr, but got an enthusiastic response from Pauli. Then I received a telephone call from Einstein, saying that he wanted to discuss the book with me. When we met, he said that I had explained Bohr's point of view as well as could probably be done, but that he was still not convinced. What came out was that he felt that the theory was incomplete, not in the sense that it failed to be the final truth about the universe, as a whole, but rather, in the sense that a watch is incomplete, if an essential part is missing. This was, of course, close to my more intuitive sense that the theory was dealing only with statistical arrays of sub-processes associated with similar observational events. Einstein felt that the statistical predictions of the quantum theory were more correct, but that by supplying the missing elements, we could in principle get beyond statistics, to, at least in principle, a determinate theory.

This encounter with Einstein had a strong affect on the direction of my research, because I then became seriously interested in whether a deterministics extension of the quantum theory could be found. In this connection, I soon thought of the classical Hamilton-Jacobi theory, which relates waves to particles in a fundamental way. Indeed, it had long been known that when one makes a certain approximation (Wentzel-Kramers-Brillouin), Schroedinger's equation becomes equivalent to the classical Hamilton-Jacobi equation. At a certain point, I suddenly asked myself: what would happen, in the demonstration of this equivalence, if we did not make this approximation? I quickly saw that there would be a new potential, representing a new kind of force, that would be acting on the particle. I called this the quantum potential, which was designated by Q.

This gave rise immediately to what I called a causal interpretation of the quantum theory. The basic assumption was that the electron *is* a particle, acted on not only by the classical potential, V, but also by the quantum potential, Q. This latter is determined by a new kind of wave that satisfies Schroedinger's equation. This wave was assumed, like the particle, to be an independent actuality that existed on its own, rather than being merely a function from which the statistical properties of phenomena could be derived. However, I showed on the basis of further physically reasonable assumptions that the intensity of this wave is proportional to the probability that a particle *actually is* in the corresponding region of space (and is not merely the probability of our observing the phenomena involved in *finding* a particle there). So the wave function had a double interpretation--first as a function from which the quantum potential could be derived, and secondly, as a function from which probabilities could be derived.

From these assumptions, one was able to show that all the usual results of the quantum theory could be obtained on the basis of a model incorporating the independent actuality of all its basic elements (field and particle), as well as an in-principle complete causal determination of the behavior of these elements in terms of all the relevant equations (at least in a one-particle system, which is as far as I had gotten at the time).

I sent pre-publication copies of this work to various physicists. De Broglie quickly sent me a reply indicating that he had proposed a similar idea at the Solvay Congress in 1927, but that Pauli had severely criticized it and that this had led him to give it up. Soon after this, I received a letter from Pauli, stating his objections in detail. These had mainly to do with the many-particle system, which I had not yet considered seriously. However, as a result of these objections, I looked at the problem again, and came out with a treatment of the many-particle system, which consistently answered Pauli's criticisms.

A more detailed consideration of this extended theory led me to look more carefully into the meaning of the quantum potential. This had a number of interesting new features. Indeed, even in the one-particle system, these showed up to some extent. For the quantum potential did not depend on the intensity of the wave associated with this electron; it depended only on the *form* of the wave. And thus, its effect could be large even when the wave had spread out by propagation across large distances. This already introduced a certain kind of *non-locality*. For example, when the wave passes through a pair of slits, the resulting interference pattern produces a complicated quantum potential that could affect the particle far from the slits, and this explains why a statistical distribution of such particles would have a pattern reflecting the wave intensity. Thus, the well known wave-particle duality of the properties of matter was explained by saying, not only that the electron is a particle that is always accompanied by a wave, but also, by noting that this wave could generally have a major effect on the particle, that reflects the whole environment. And in this way, one was able to obtain a further insight into the crucially significant new feature of wholeness of the electron and its environment which Bohr had shown to be implicit in the quantum theory.

When one looked at the many-particle system, this new kind of wholeness became much more evident. For the quantum potential was now a function of the positions of all the particles which (as in the one-particle case) did not necessarily fall off with the distance. Thus, one could at least in principle have a strong and direct (non-local) connection between particles that are quite distant from each other. What was more striking, however, was that the very form of this connection depends on the wave function for the state of the whole. This is determined by solving Schroedinger's equation for the entire system, and thus does not depend on the state of the parts. Such a behavior is in contrast to that shown in classical physics, for which the interaction between the parts is a predetermined function, independent of the state of the whole. Thus, classically, the whole is merely the result of the parts and their pre-assigned interactions, so that the primary reality is the set of parts while the behavior of the whole is derived entirely from those parts and their interactions. With the quantum potential, however, the whole has an independent and prior significance, such that, indeed, the whole may be said to organize the parts. In a certain sense, quantum wholeness is thus closer to the organized unity of a living being that it is to that obtained by putting together the parts of a machine.

If the whole is a primary notion in quantum mechanics, how do we account for our usual experience of a world made up of a vast set of independent elements that can correctly be understood in terms of ordinary mechanical notions? The answer is that when the wave function of the whole system reduces to a set of constituent factors, the quantum potential simplifies to a sum of independent components. As a result, the whole reduces to a set of independent sub-wholes. One can show that this is the situation that will commonly prevail at the large-scale level. But more generally, the wave function does not factorize, and the whole cannot be divided into such independent sub-wholes.

To sum up, then, the quantum potential not only organizes wholes; it determines which sub-wholes, if any, that may be within the whole. It is clear how radically new are these implications of the quantum theory. They are hinted at only vaguely and indirectly by the subtle arguments of Bohr, based on the usual interpretation of the quantum theory as nothing more than a set of mathematical formulae yielding statistical predictions of the phenomena that are to be obtained in physical observations. However, by putting quantum and classical theories in terms of the same intuitively understandable concepts (particles moving continuously under the action of potentials), one is able to obtain a clear and sharp perception of how the two theories differ. I felt that such an insight was important in itself, even if, as seemed likely, this particular model could not provide the basis for a definitive theory that could undergo a sustained development. For a clear intuitive understanding of the meaning of one's ideas can often be helpful in providing a basis from which may ultimately come an entirely new set of ideas, dealing with the same content.

These proposals did not actually "catch on" among physicists. The reasons are quite complex and difficult to assess. Perhaps the main objection was that the theory gave exactly the same predictions for all experimental results as does the usual theory. I myself did not give much weight to these objections. Indeed, it occurred to me that if de Broglie's ideas had won the day at the Solvay Congress of 1927, they might have become accepted interpretation. Then if someone had come along to propose the current interpretation, one could equally well have said that, since, after all, it gave no new experimental results, there would be no point in considering it seriously. In other words, I felt that the adoption of the current interpretation was a somewhat fortuitous affair, since it was affected not only by the outcome of the Solvay Conference, but also by the generally positivist empiricist attitude that pervaded physics at the time. This attitude is in many ways even stronger today, and shows up in the fact that a model that gives insight without an "empirical pay-off" cannot be taken seriously.

I did try to answer these criticisms to some extent, by pointing out that the enriched conceptual structure of the causal interpretation was capable of modifications and new lines of development that are not possible in the usual interpretation. These could, in principle, lead to new empirical predictions. But, unfortunately, there was no clear indication of how to choose such modifications, among the vast range that was possible. And so, these arguments had little effect as an answer to those who require a fairly clear prospect of an empirical test before they will consider an idea seriously.

In addition, it was important that the whole idea did not appeal to Einstein, mainly because it involved the new feature that all connections had to be local. I felt this response of Einstein was particularly unfortunate, as it almost certainly "put-off" some of those who might otherwise have been interested in this approach. Nonetheless, I saw clearly at the time that the insight that it afforded was an important reason why it should be considered, at least as a supplement to the usual interpretation. To have some kind of intuitive model was better, in my view, than to have none at all. For without such a model, research in the quantum theory will consist mainly of the working out of formulae and comparing these calculated results with those of experiment. Even more important, the teaching of quantum mechanics will reduce (as it has in fact tended to do) to a kind of indoctrination, aimed at fostering the belief that such a procedure is all that is possible in physics. Thus, new generations of students have grown up who are predisposed to consider such questions with rather closed minds.

Because the response to these ideas was so limited, and because I did not see clearly, at the time, how to proceed further, my interests began to turn in other directions. During the sixties, I began to direct my attention toward *order*, partly as a result of a long correspondence with an American artist, Charles Biederman, who was deeply concerned with this question. And then, through working with a student, Donald Schumacher, I became strongly interested in *language*. These two interests led to a paper on order in physics, and on its description through language. In this paper, I compared and contrasted relativistic and quantum notions of order, leading to the conclusion that they contradicted each other, and that new notions of order were needed.

Being thus alerted to the importance of order, I saw a program on B.B.C. television, showing a device in which an ink drop was spread out through a cylinder of glycerine, and then brought back together again, to be reconstituted essentially as it was before. This immediately struck me as very relevant to the question of order, since, when the ink drop was spread out, it still had a "hidden" (i.e., non-manifest) order, that was revealed when it was reconstituted. On the other hand, in our usual language, we would say that the ink was in a state of "disorder" when it was diffused through the glycerine. This led me to see that new notions of order must be involved here.

Shortly afterwards, I began to reflect on the hologram, and to see that in it, the entire order of an object is contained in an interference pattern of light that does not appear to have such an order at all. Suddenly, I was struck by the similarity of the hologram and the behavior of the ink drop; I saw that what they had in common was that an order was *enfolded*. That is, in any small region of space, there may be "information" which is the result of enfolding an extended order, and which could then be unfolded into the original order (as the points of contact made by the folds in a sheet of paper may contain the essential relationships of the total pattern displayed when the sheet is unfolded).

Then, when I thought of the mathematical form of the quantum theory (with its matrix operation and Green's functions), I perceived that this too described just a movement of enfoldment and unfoldment of the wave function. So the thought occurred to me: perhaps the movement of enfoldment and unfoldment is universal, while the extended and separate forms that we commonly see in experience are relatively stable and independent patterns, maintained by a constant underlying movement of enfoldment and unfoldment. This latter I called the *holomovement*. The proposal was thus a reversal of the usual idea. Instead of supposing that extended matter and its movement are fundamental, while enfoldment and unfoldment are explained as a particular case of this, we are saying that the implicate order of the holomovement is fundamental, and that the explicate order of extended and separate forms is only a particularly distinguished case of the implicate order, which is derived from the latter by unfoldment.

This approach implies, of course, that each extended form is enfolded in the whole, and that, the whole is enfolded in this form (though, of course, there is an asymmetry, in that the form enfolds the whole only in a limited and not completely defined way). The way in which the extended form enfolds the whole is however not merely superficial or of secondary significance, but rather, it is essential to what that form *is* and to how it acts, moves, and behaves quite generally. So the whole is, in a deep sense, *internally* related to the parts. And, since the whole unfolds all the parts, these latter are also internally related, though in a weaker way than they are related to the whole.

I shall not go into great detail about the implicate order here; I shall assume that most readers are somewhat familiar with this. What I want to emphasize is only that the implicate order provided an image, a kind of metaphor, for intuitively understanding that implication of wholeness which is the most important new feature of the quantum theory. Nevertheless, it must be pointed out that the specific analogies of the ink drop and the hologram are limited, and do not fully convey all that is meant by the implicate order. What is missing is the fact that the parts or sub-wholes not only unfold from the whole; but they unfold in a self-organizing and stable way. On the other hand, in both these models, there is no inner principle of organization that determines the parts of sub-wholes and makes them stable. In fact, the order enfolded in the whole is obtained from pre-existent, separate and extended elements (objects photographed in the hologram or ink drops injected into the glycerine). It is then unfolded to give these elements again. Nor is there any natural stability in these elements; they may be totally altered or destroyed by minor further disturbances of the over-all arrangement of the equipment.

Gradually, throughout the seventies, I became more aware of the limitations of the hologram and ink droplet analogies to the implicate order. Meanwhile, I noticed that both the implicate order and the causal interpretations had emphasized this wholeness signified by quantum laws, though in apparently very different ways. So I wondered if these two rather different approaches were not related in some deep sense--especially because I had come at least to the essence of both notions at almost the same time. At first sight, the causal interpretation seemed to be a step backwards toward mechanism, since it introduced the notion of a particle acted on by a potential. Nevertheless, as I have already pointed out, its implication that the whole both determines its sub-wholes and organizes them clearly goes far beyond what appeared to be the original mechanical point of departure. Would it not be possible to drop this mechanical starting point altogether?

I saw that this could indeed be done by going on from the quantum mechanical particle theory to the quantum mechanical field theory. This is accomplished by starting with the classical notion of a continuous field (e.g., the electromagnetic) that is spread out through all space. One then applies the rules of the quantum theory to this field. The result is that the field will have discrete "quantized" values for certain properties, such as energy, momentum, and angular momentum. Such a field will act, in many ways, like a collection of particles, while at the same time, it still has wave-like manifestations, such as interference, diffractions, and so on.

Of course, in the usual interpretation of the theory, there is no way to understand how this comes about. One can only use the mathematical formalism to calculate statistically the distribution of phenomena through which such a field reveals itself in our observations and experiments. But now, one can extend this causal interpretation to the quantum field theory. Here, the actuality will be the entire field over the whole universe. Classically, this is determined as a continuous solution of some kind of field equation (e.g., Maxwell's equations for the electromagnetic field). But when we extend the notion of the causal interpretation to the field theory, we find that these equations are modified by the action of what I called a super-quantum potential. This is related to the activity of the entire field as the original quantum potential was to that of the particles. As a result, the field equations are modified, in a way that make them, in technical language, non-local and non-linear.

What this implies for the present context can be seen by considering that, classically, solutions of the field equations represent waves that spread out and diffuse independently. Thus, as I indicated earlier in connection with the hologram, there is no way to explain the origination of the waves that converge to a region where particle-like manifestion is actually detected, nor is there any factor that could explain the stability and sustained existence of such a particle-like manifestation. However, this lack is just what is supplied by the super-quantum potential. Indeed, as can be shown by a detailed analysis, the non-local features of this latter will introduce the required tendency of waves to converge at appropriate places, while the non-linearity will provide for the stability of recurrence in the whole process.

Out of this emerges a picture of a wave which spreads out, and converges, again and again, to show a kind of average particle-like behavior, while the inference and diffraction properties, of course, are still maintained. All this flows out of the activity of the super-quantum potential, which depends in principle on the state of the whole universe. But if the "wave function of the universe" falls into a set of independent factors, at least approximately, a corresponding set of relatively autonomous and independent sub-units of field function will emerge. So now, we see that the whole universe not only determines and organizes its sub-wholes; it also gives form to what have until now been called the elementary particles out of which everything is supposed to be constituted. What we have here is a kind of universal process of constant creation and annihilation arranged into a world of form and structure, in which all manifest features are only relatively constant, recurrent, and stable aspects of the whole.

To see how this is connected with the implicate order, we have only to note that the original holographic model was one in which the whole was constantly enfolded into and unfolded from each region of an electromagnetic field, through dynamical movement and development of the field according to the laws of classical field theory. But now, this whole field is no longer a self-contained totality; it depends crucially on the super-quantum potential. When one looks at the mathematical form of the latter, one discovers that this too is a kind of implicate order. But it is immensely more subtle than that of the original field, as well as more inclusive in the sense that not only is the actuality of the field enfolded in it, but also, all the possibilities.

I was therefore led to call the original field the first implicate order, while the super-quantum potential was called the second implicate order (or the super-implicate order). In principle, of course, there could be a third, fourth, fifth implicate order, going on to infinity, and these would correspond to extensions of the laws of physics going beyond those of the current quantum theory, in a fundamental way. But for the present, I want to consider only the second implicate order, and to emphasize that this stands in relationship to the first as a source of formative, organizing, and creative activity.

It should be clear that this notion now incorporates both of my earlier perceptions--the implicate order as a movement of outgoing and incoming waves, and of the casual interpretation of the quantum theory. So, although these two ideas seemed initially very different, they proved to be two aspects of one more comprehensive notion. This can be described as an over-all implicate order, which may extend to an infinite number of levels, and which differentiates and organizes itself into independent sub-wholes, while determining how these are inter-related to make up the whole.

It is possible, moreover, that the principles of organization of such an implicate order may even define a unique explicate order, as a particular and distinguished sub-order, in which all the elements are relatively independent and externally related. Or to put it differently, the explicate order itself may be obtainable from the implicate order as a special and determinate sub-order that is contained within it.

All that has been discussed here opens up the possibility of considering the cosmos as an unbroken whole through an over-all implicate order. Of course, this possibility has been studied thus far in only a preliminary way, and a great deal more work is required to clarify and extend the notions that have been discussed in this paper.

Birkbeck College
University of London